From Crystals to Kites

Exploring Three Dimensions

Ron Kremer
Dale Seymour Publications

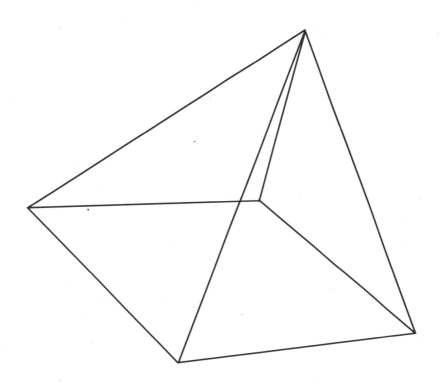

Project Editor: Joan Gideon
Production/Mfg Coordinator: Leanne Collins
Design Manager: Jeff Kelly
Text and Cover Design: Dagen Bela
Cover Illustration: Thom Ricks
Text Illustration: Ron Kremer
Special thanks to *National Geographic Magazine* for article use.

Published by Dale Seymour Publications, an imprint of Addison-Wesley's Alternative Publishing Group.

Order Number DS21328
ISBN 0-86651-798-7

3 4 5 6 7 8 9 10-ML-99 98 97 96 95

This book is printed
on recycled paper.

Contents

Introduction

From Crystals to Kites: Exploring Three Dimensions is a series of explorations that begins with observations of crystals in geology and ends with analyzing models of solids in geometry. As you use this book to explore and discover with your students, you will encounter applications in math and science. Your students will learn by doing.

Actively involved in open-ended experimentation, students observe, experiment, and evaluate. As they organize data they have collected, students look for patterns and trends, making generalizations that can be tested by extrapolation and further experimentation. The explorations are more effective if presented sequentially, although most of them can be used independently.

In this exploratory climate, you are the facilitator, resource person, and encourager. Your students are active participants. Allow them to explore without giving hints that might cut short the discovery process. Instead, suggest strategies that would lead to a solution only after it is obvious that the whole class is stuck. Allow students to post incorrect data during collection time. When the group is in the process of organizing the posted data, errors become obvious because they don't fit the emerging pattern. If they don't discover the error themselves during an evaluation session, their peers will. If no one finds a posted error, then point it out.

Use short periods over several days rather than accomplishing everything in one day. These explorations are most effective if students have time between steps in the process to let previous experiences sink in. Throughout the day, students may look at the posted data and discover an error, a missing solution, or a new pattern.

Be patient! The best way for a student to learn mathematics is through interactive, hands-on explorations. These take more time than traditional teacher-directed lessons. There will be more noise as well. You will hear the productive sounds of materials being manipulated and the natural verbal interaction of peers solving problems together. But the extra commotion is worth the excitement of discovery when the cognitive domain is altered, concepts are restructured, and students' personal experience grows. This kind of excitement caused the discoverer of the relationship between water displacement, buoyancy, and density to leap from his tub and run dripping into the street yelling, "Eureka!" The act of discovery causes excitement. Even one discovery may change a child's perception of the world for the rest of his or her life!

Getting Organized

In an exploratory, hands-on classroom, manipulatives must be available for each student. For many explorations in this book, each student or pair of students can have their own materials or manipulatives. For other explorations this may be too expensive. These activities can be done in small groups. If you divide the class into four groups, each group can work on a different activity, and you will need fewer materials. A group of eight students working in pairs will only need four sets of expensive materials.

Once the exploration has been introduced to the whole class, one group can begin working with the materials—exploring, building, and gathering and recording data. You can continue to work with the rest of the class as a group or have all students working in groups on other math-related activities, such as review and practice games, textbook activities, or writing in their math journal.

Whatever classroom organization works for an exploration, provide an area where students can display their results. When all groups have had a chance to work with the materials, the class as a whole can participate in identifying patterns, drawing conclusions, and assessing their work.

A First Activity

Describing Behaviors of Scientists

Students will discuss what scientists do and suggest verbs to describe those activities.

Materials
Long strips of heavy paper
Markers

Exploration
In a brainstorming session with your students, describe the activities of astronomers, zoologists, botanists, geologists, and other scientists. Ask students to suggest verbs that describe these activities. Individually or in groups, students can write these verbs on strips of paper.

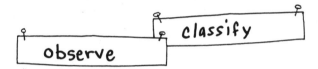

Closing Discussion
After students have had time to write several verbs, post the strips, removing duplicate verbs. With your students, talk about the posted verbs, make sure each verb describes the behavior of a scientist. If students do not include *read, write, study,* and *think* in their list of verbs, add them and discuss how they are a part of scientific behavior. The verbs your students suggest and their comments will help you understand their perception of science and scientists' activities.

When the students are satisfied that the list is accurate and fairly complete, ask the class which verbs described their own behavior during this activity. Some of their responses may be *record, collect, post,* and *evaluate.*

Point out that the verbs they have posted and talked about describe what they will be doing as they participate in the explorations. Save these verb strips and hold up the appropriate ones during some of the explorations to remind the students of how they are behaving like scientists.

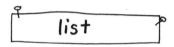

Career Awareness
Have students investigate careers in astronomy, zoology, botany, geology, and other sciences.

Exploration 1
Exploring Crystals

Students will observe crystals with simple microscopes, make drawings, and record descriptive comments.

Materials
Several simple microscopes or hand lenses
See page 7 for directions on building these.
Copies of Using the Microscope (page 9)
Crystals for student investigation
Three or four kinds of crystals such as
* *Table salt (Be sure that moisture hasn't rounded the corners.)*
* *Sugar (Use the large, uncolored crystals used by bakeries to sprinkle on baked goods.)*
* *Copper sulfate (This bright-blue crystal is excellent to observe, but it is poisonous.* **Warning:** *Caution students not to taste the sample or get it in their eyes. Tell them it is poisonous.)*
* *Epsom salts*
* *Sodium citrate*

Small bottles and bags for crystal samples
Clean the bottles thoroughly. Number the lids, making a set of bottles for each microscope. Put small samples of the crystals into the bottles always using the same bottle number for the same crystal so students can compare their data.

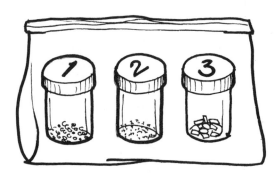

Put each set of bottles into a resealable plastic bag.
Large crystals for demonstration
Use samples such as quartz and pyrite. Granite containing large and small crystals shows crystals in the context of a rock.
Copies of Data Collection Form (page 8)

Introduction
Show some sample crystals and granite or other rocks that have crystals in them. The large, dark crystals in granite are hornblende; the lightest-colored crystals are quartz. If you have different samples of granite showing large crystals in one and small crystals in the other, talk with students about why the crystals are different sizes. The larger the crystals, the slower the sample cooled.

Tell the students they are going to observe, draw, and describe several samples of crystals. Discuss the definition of a crystal: A shape with a regularly repeating internal arrangement of atoms and external boundaries or faces formed as a mineral solidifies.

Exploration
Pass out copies of Using the Microscope, and demonstrate how to set up and use the microscope.

Pass out copies of the Data Recording Form and read it over with the class.

Tell the students not to touch the crystals as the moisture on their hands will ruin the crystalline structure. Students should tap a few crystals onto the viewing stage. After each crystal is viewed, have students remove the tripod magnifier, tilt the stage over the bottle, and tap the edge. If you are using the plastic petri dishes instead of the Plexiglas stage, students can pick up the dish and dump the contents back into the correct bottle. Don't pick up spilled crystals and put them back in the bottle.

Remind the students not to mix the crystals.

If students are using copper sulfate, tell them it is an eye irritant and poisonous. They would need to eat a lot of it to become sick, but caution them not to taste it.

Distribute the microscopes and crystals, and allow the students to begin the observations. Encourage them to draw two or three individual crystals instead of dumping the entire contents of the bottle onto the stage and attempting to draw a mass of crystals. Emphasize the importance of keeping neat and complete records.

Closing Discussion

When all students have had an opportunity to complete the crystal observations, ask them to share some of their observations. As the students share, record their data on the board so other students can compare their findings. You may wish to ask students to make their own drawings on the board.

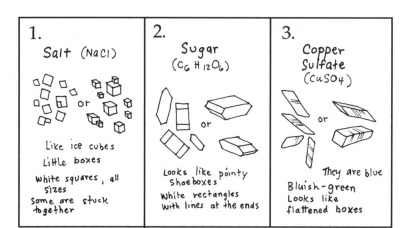

1. Salt (NaCl)

Like ice cubes
Little boxes
White squares, all sizes
Some are stuck together

2. Sugar (C₆H₁₂O₆)

Looks like pointy shoeboxes
White rectangles with lines at the ends

3. Copper Sulfate (CuSO₄)

They are blue
Bluish-green
Looks like flattened boxes

Ask students to describe the crystals. If a student says that a salt crystal is like an ice cube, use the opportunity to point out that this form of crystal is called a *cube*. If no one uses *cube*, tell the class that the salt crystals are cubes.

Sugar is a prism with two faces that are hexagons, or truncated parallelograms. All of its other faces are rectangles.

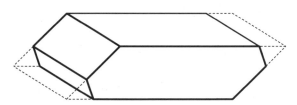

Remind students of the different sizes of similar crystals in the two samples of granite you showed during the introduction. Ask them why sugar crystals from the bakery are larger than the sugar crystals they use at home. At the sugar production plant some crystals were given more time to grow, making them larger.

A copper sulfate crystal has three pairs of parallel parallelograms and is therefore called a *parallelepiped*. A copper sulfate crystal has no right angles. Imagine a shoe box being slightly flattened in one direction and then being slightly flattened in a second direction at a right angle to the first.

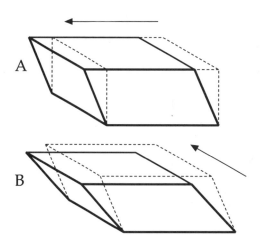

Geologists call this a *triclinic crystal*. Nose of the axes are perpendicular to each other.

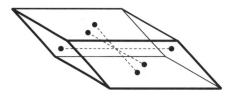

In a cube, all axes are perpendicular.

Tell students to save their Data Collection Form for reference in Exploration 2.

Career Awareness

Invite a geologist to your classroom to share and explain samples of crystals. In preparation for this visit have students develop a list of questions.

Independent Investigation

Some students may really get excited about crystallography. Encourage them to find books in the library about growing their own crystals. Some science stores sell crystal kits from which large crystals can be grown easily. Students can also grow crystals using the brine water from an ice cream maker. Pour it into a plastic dishpan to a depth of about one inch. Allow it to evaporate slowly. After several days, large, flat squares of salt crystals will form. Some will be up to one-half inch on a side. The growth pattern can easily be seen. Encourage students who grow crystals to bring them to class.

Directions for Building a Simple Microscope

Materials

Four $1\frac{1}{2}$" nails

Two pieces of $\frac{3}{4}$" wood, 3" × 3"

A piece of $\frac{3}{4}$" wood, 3" × 4"

A piece of $\frac{1}{8}$" clear rigid plastic, 3" × 3"

Heavy tin foil plate

Flat black paint

Wire coat hanger

Wood glue

Tripod 10X (ten-power) magnifier

Procedure

1. Cut a groove $\frac{1}{4}$" deep, $\frac{1}{4}$" down from the top of each 3" × 3" piece of wood.

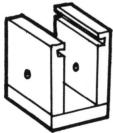

2. Drill $\frac{1}{8}$" diameter hole $\frac{1}{2}$" below the center of each 3" × 3" piece.

3. Nail both 3" × 3" pieces to 3" × 4" base with the grooves facing the inside. (Lightly apply wood glue before nailing.)

4. Paint with flat black to provide a dark background for viewing.

5. Straighten and cut 7" of coat hanger wire. Push the wire through both holes and bend down 1" on one side to form a right angle. On the other side, bend the remaining wire down slightly for a handle.

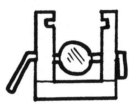

6. Cut a 2" diameter mirror from the bottom of a foil plate and glue it to the center of the wire.

7. Slide the plastic into grooves and set tripod magnifier on top. (Instead of a plastic stage you may want to use a plastic petri dish. It is better for observing water samples.)

Data Collection Form: OBSERVING CRYSTALS

Name _____

- **What you need to begin**
  Samples of crystals in bottles
  Microscope or 10X (ten-power) lens

- **Before you begin**
  Please do not mix the crystals. Open
  only one bottle at a time.

  Using just a few crystals makes
  them easier to see and to draw.

  If you pick up the crystals with your
  fingers, you will ruin them.

  Caution: One of the crystals is
  poisonous! Don't taste it or get it
  into your eyes.

- **During the observation**
  Look carefully at each crystal
  sample and draw several separate
  crystals. Also use words to describe
  the crystals.

- **After the observation**
  Please brush spilled crystals into the
  wastebasket. Put all the bottles back
  into the bag. Leave the tripod
  magnifier on the wooden stand.

1.

2.

3.

Using the Microscope

Plastic stage

Petri dish stage

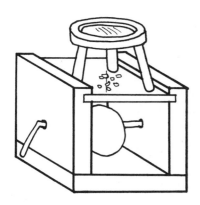

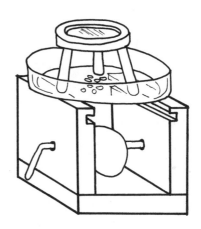

Sprinkle a few crystals on top of the plastic.

Carefully set the tripod magnifier on the plastic over the crystals.

Use the handle to turn the mirror. Adjust the mirror to change the amount of light for best viewing.

To view crystals against a black background, turn the mirror straight up and down.

Set the petri dish on top of the wooden frame.

Sprinkle a few crystals into the center of the petri dish.

Carefully set the tripod magnifier in the petri dish over the crystals.

Use the handle to turn the mirror. Adjust the mirror to change the amount of light for best viewing.

To view the crystals against a black background, turn the mirror straight up and down.

Exploration 2
Building Soda Straw Models

Students will build models of idealized crystals and other geometric solids.

Materials
Ten plastic soda straws for each student
Transparent tape
> *Three or four students may share a roll of tape, or give each student a long piece of tape to cut smaller pieces from.*

Scissors
Data Recording Forms from Exploration 1

Introduction
Explain to the students that they will use straws and transparent tape to build models of crystals. Demonstrate how to use the straws and transparent tape to build models. Place tape along the length of the straws and roll it around to make a good connection.

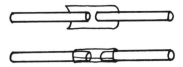

Leave some space between the straws to keep them from being pinched or deformed when the connection is bent.

Students should cut the straws into pieces rather than using whole straws. Fold the straw and crimp the fold. Cut on the fold and pop the ends of the straw open again.

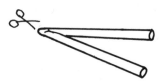

Save time by bending straws into plane figures and taping them. Then combine plane figures to build models of the crystals.

To build a prism, make a pair of congruent figures and connect them with straws of equal lengths.

If students share rolls of tape, encourage them to tear off a number of short pieces and lightly tack them to the edge of their desks while another student is cutting pieces of straw.

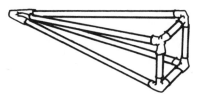

Exploration
Distribute the straws, transparent tape, and scissors. In a few instances a student may need to use whole straws. But limit the use of whole straws by limiting the number of straws each student receives.

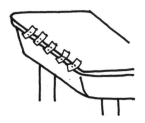

Remind students to refer to the drawings they made in Exploration 1 for ideas of crystal shapes. Tell students that models do not have to look exactly like the crystals. In fact, encourage them to try to build other forms also.

As you move around the room, watch for students who are adding unnecessary straws to their constructions. Diagonals only complicate the model of a solid. Explain the concept of a diagonal and diagram one on the board.

Closing Discussion
When all students have completed their models, evaluate them. Some students may have constructed several models. Tell students to check the tape connections, especially any loose vertices or wobbly models. Loose vertices can be tightened up

by squeezing the tape connections. Show several completed models, finding some models that have unnecessary straws or diagonals as examples. Be sure to use models by students who are comfortable having their work criticized. Ask whether their models have any unnecessary straws or diagonals. Students working in small groups of two or three can check each other's models.

To save the models for the next exploration, ask students to write their names on pieces of masking tape, tag their models, and place them on a table or counter top.

Independent Investigation

After the model-building session is over, some students may be reluctant to quit. Encourage them to build models at home and bring them to school to show and evaluate. Include these models in the next activity.

Exploration 3
Comparing Soda Straw Models

Students will count edges, faces, and vertices of solids, looking for the pattern that relates them. They will learn correct terms to identify solids, and compare characteristics of solids.

Materials
Transparent tape
Soda straw model of a rectangular prism
White crayons
Black, or other dark-colored construction paper
    (9" × 12")
Several large pieces of butcher paper for charts
Charts
    Prisms and Pyramids (page 16)
    Polyhedra (page 17)
Brightly colored paper
    Cut out the letters and symbols of the equation
    $F + V - 2 = E$ *for display on the bulletin board.*

Introduction
Explain that students will use their straw models to explore geometric solids.

Show a large model of a rectangular prism. Explain that each piece of straw is called an *edge*. Ask how many edges this straw model has. Help students count edges if necessary. Show four on the top, four on the bottom, and four connecting the top and the bottom for a total of twelve edges. Explain that a vertex is where two or more straws (edges) meet at a corner. Count the number of vertices this model has—four on the top and four on the bottom for a total of eight. Explain that every area bounded by edges is called a face. In this case, the faces are rectangles. Determine how many faces the rectangular prism has by counting four around the sides and including the top and bottom for a total of six. On the board, make a list like this.

Edges = 12
Vertices = 8
Face = 6

Exploration
After counting the edges, faces, and vertices of their straw models, students can record the data. This data will be evaluated later, but students may refer to it during the discussion and revise it if errors are discovered in the process of sharing.

Introduce the correct term for each solid as the appropriate models are explored. See pages 16–17, and 22 for charts of solids.

Show one of the models and ask students to find a similar model. Have the class evaluate the responses. For example, if you show the rectangular prism they might respond:

Ask how their models are like yours. They are all rectangular prisms, with six faces, eight vertices, and twelve edges. If someone shows a shape that doesn't fit the groups' description, ask that person what their rule is. Why does it belong with this set of solids? If it does fit, the group may need to revise their definition of that set.

These models are also somewhat like a rectangular prism.

Discuss how these are all alike. They are prisms, each have one pair of congruent faces as bases, and all faces that connect them are parallelograms. This definition includes any prism, including the rectangular prism mentioned earlier. See page 16 for a chart illustrating prisms. Some students may notice that prisms don't all have the same number of faces, vertices, and edges.

If one face has a diagonal, it must be counted as an edge, and the face must be counted as two faces. Students who love using tape may add many unnecessary diagonals making the model more complicated to evaluate. One common shape, a house, is often built with a diagonal across the top of the square; encourage students to simplify it to a pentagonal prism.

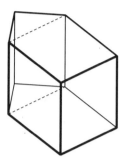

Show one of the larger models of a pyramid and ask students to hold up similar models. Some responses might be:

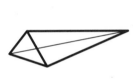

Ask how these are alike. They are pyramids, with a polygon for a base and triangles that meet at a common vertex for the remaining faces. See page 16 for a chart illustrating pyramids.

There are polyhedra that are not prisms or pyramids. Page 17 illustrates some of these. Students may add to pyramids by turning the base into another pyramid creating a hexahedron or octahedron. They may use each face of a prism as the base of a pyramid creating polyhedra like the extended tetrahedron and extended cube pictured on page 17. Students may begin with a hexagon and add squares and triangles to make a cuboctahedron. Truncated polyhedra can be built; occasionally after a model is completed students will cut and revise each vertex. Use the drawings to help students name their models or to give them ideas of polyhedra to build. For more complicated or irregular solids, make available a good geometry resource book, such as *Mathematics: A Human Endeavor*, 2nd ed. by Harold R. Jacobs (San Francisco: W. H. Freeman, 1982), or *Polyhedron*

Models by Magnus J. Wenniger (Cambridge: Cambridge University Press, 1971).

When students have completed their straw models, distribute pieces of black construction paper, transparent tape, and white crayons. Ask student to tape their straw model to a piece of black construction paper, then, using a white crayon, write the number of faces, vertices, and edges below the model.

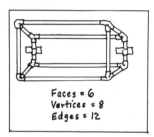

Faces = 6
Vertices = 8
Edges = 12

Display completed models with data lists. Your students can group these by prisms, pyramids, and other polyhedra.

To look for patterns in this data, draw the following charts on the board or on large pieces of butcher paper.

Prism	Faces	Vertices	Edges

Prism	Faces	Vertices	Edges
	5	6	9
	6	8	12
	7	10	15
	8	12	18
	9	14	21
	10	16	24

Begin with the prisms. Students can use information posted on the bulletin board to complete the first row. Ask which prism comes next on the chart. Draw in the rectangular prism. Then ask them to complete the second row with the data on the rectangular, or square, prism. Continue this for pentagonal and hexagonal prisms. There may not be any models to refer to, but encourage the students to look for a pattern in the columns to help them extrapolate to find the last several rows. Students can build models to confirm their extrapolation or count the edges and vertices on drawings of these shapes.

When the chart is completed, the class can find other patterns. The vertical patterns down the columns are easiest to see. The prisms are arranged so each has one more face than previous one. In the column of vertices, the numbers increase by two. For edges, add three each time.

There is also a horizontal relationship between the faces and the vertices—as you move down, the number you add to the faces to equal the vertices, n, increases by one each time the number of faces increases.

faces + n = vertices

$5 + 1 = 6$

$6 + 2 = 8$

$7 + 3 = 10$

There is a similar relationship between the vertices and the edges.

vertices + n = edges

$6 + 3 = 9$

$8 + 4 = 12$

$10 + 5 = 15$

Follow the same procedure with the pyramids. Students can provide data on the triangular pyramid in the first row, referring to the bulletin board.

Pyramid	Faces	Vertices	Edges

Pyramid	Faces	Vertices	Edges
	4	4	6
	5	5	8
	6	6	10
	7	7	12
	8	8	14

Ask them which pyramid comes next on the chart. Draw in the square pyramid. Have them complete the data for the square pyramid. Do the same for the pentagonal pyramid. There may not be any student models to refer to, but they can look for patterns in the columns and extrapolate or count edges and vertices on a drawing to complete the chart. Ask students what other patterns they see in the completed chart of pyramids. Patterns down the columns are usually found first, but because pyramids have the same number of faces and vertices, that relationship across the rows is recognized easily.

Eventually someone will notice that adding the number of faces and vertices of any prism or pyramid, then subtracting two from the sum always equals the number of edges. The student who discovers this relationship has duplicated the work of Leonhard Euler (pronounced "oiler") who noted the relationship between faces, vertices, and edges of a polyhedron over two hundred years ago.

Post the equation, $F + V - 2 = E$, on the bulletin board with the straw models. Students will refer to the equation as they explore other polyhedra.

Closing Discussion

Encourage students to use Euler's equation to test all the posted data. Data that are incorrect need to be corrected, then reposted.

Independent Investigation

Some students will not be satisfied until they have thoroughly tested Euler's equation with more complex straw models.

Prisms and Pyramids

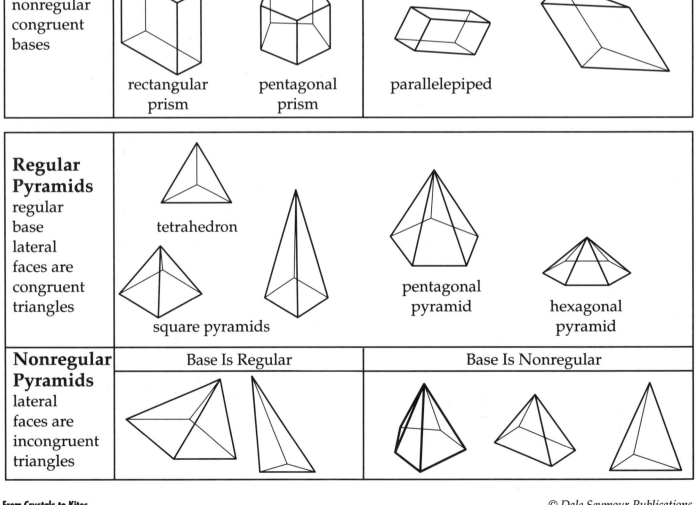

	Right Prisms	**Oblique Prisms**
Regular Prisms regular congruent bases	triangular prism square prism hexagonal prism cube	
Nonregular Prisms nonregular congruent bases	rectangular prism pentagonal prism	parallelepiped

Regular Pyramids regular base lateral faces are congruent triangles	tetrahedron square pyramids	pentagonal pyramid hexagonal pyramid
Nonregular Pyramids lateral faces are incongruent triangles	Base Is Regular	Base Is Nonregular

Polyhedra

Regular Polyhedra

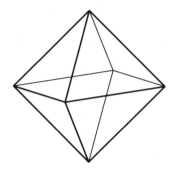

octahedron

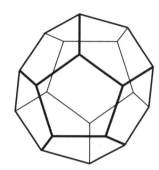

dodecahedron

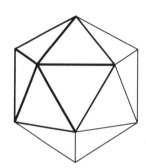

icosohedron

Semiregular and Nonregular Polyhedra

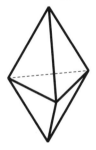

hexahedron

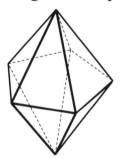

decahedron

dodecahedron

cuboctahedron

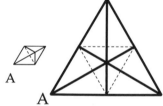

extended tetrahedron

extended cube

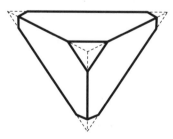

truncated tetrahedron

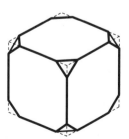

truncated hexahedron

Exploration 4
Exploring Polyhedra with Commercial Construction Sets

Students will construct polyhedra with a variety of commercial materials, use correct terminology to identify polyhedra, and compare characteristics of polyhedra. These construction sets allow the students to build much more complex polyhedra. Set angles and lengths of edges often result in regular polyhedra.

Materials
Ramagon
> *Tinkertoys may be substituted for Ramagon.*

Polydrons
> *Teacher-made paper polygons may be substituted for Polydrons.*
>> *Copy the patterns on pages 23–24 on heavy paper, laminate them, cut them out, punch holes at each vertex to the fold lines, crease the flaps, use small rubber bands to hold the piece together. The last edge will need to be tucked in.*

Ramagon prism and drawing
> *Construct and draw this prism, use the drawing to demonstrate how to record Ramagon solutions.*

Polyhedra charts (pages 16, 17, and 22)
> *Use these drawings as records of complex polyhedra. Enlarge the drawings on a copy machine first, then cut out and laminate the copies.*

Tetrahedron built from Polydrons and one loose square

White art paper

Masking tape
> *Use the masking tape to mark edges or vertices when counting.*

Ramagon

Ramagon is a set of plastic balls and rods that become the vertices and edges of polyhedra. Like the straw models, the edges and vertices define the polyhedra.

Introduction

Explain to the students that they will construct polyhedra using Ramagon (or Tinkertoys) and record their solutions by drawing what they have built. Show the class the prism and its two-dimensional drawing. Demonstrate the process of making a drawing from a prism.

1. Trace the base of the prism.

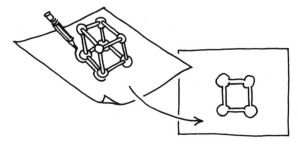

2. Move the prism a diagonal distance equivalent to its height and trace the base again.

3. Draw lines connecting the vertices of the two bases.

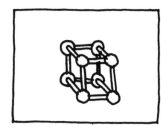

All the prism solutions can be recorded in this way. The tracing method will work for making top views of pyramids also.

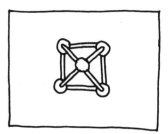

For other constructions students may have to settle for showing the polyhedron before taking them apart. Or, if the polyhedron is one shown on pages 16, 17, or 22, the student can post an enlarged copy of the drawing.

Show how to use the Ramagon connector balls and rods. Squeeze the slotted end of a rod and push it into a hole on the ball connector and pull on it to show that it is locked; squeeze the end of the rod again to pull it out of the hole. Ask the students what a ball represents and what a rod represents in a polyhedron model. (A ball represents a vertex; a rod represents an edge.)

Exploration

After students have had a chance to explore the materials and build several polyhedra, let them find further ideas from the project booklet that comes with the construction set. As each polyhedron is finished, have students count faces, vertices, and edges and test the data against Euler's equation.

Closing Discussion

When all students have explored Ramagon's possibilities, ask them to share their polyhedra record. Review the names of polyhedra that students have built and then taken apart, referring to drawings that have been posted.

Some of the polyhedra that are pictured cannot be made with Ramagon—icosahedron, dodecahedron, icosidodecahedron, rhombicosidodecahedron, great rhombicosidodecahedron, snub cuboctahedron, and snub dodecahedron. Explore with the class the characteristics of these polyhedra—they all have pentagons or arrangements of five triangles; they are all in the icosa- or dodeca- family.

Ask students to count faces, vertices, and edges for these polyhedra. If any of them seem to not fit Euler's equation, have several students recount the polyhedron's faces, vertices, and edges.

Have the students group the posted polyhedra. Depending on the data base, the students will find at least three groups—prisms, pyramids, and a mixed group, which they will learn to sort later. Students can check the characteristics of the polyhedra in each group to be sure they are sorted correctly.

Polydrons

Polydrons, a set of interlocking plastic triangles, squares, pentagons, and hexagons, is the inverse of Ramagon. In Ramagon, you build with edges and vertices, creating faces. In Polydrons you build with faces, creating edges and vertices.

Introduction

Explain to the students that they are going to be constructing polyhedra using Polydrons.

Show the class the tetrahedron you built from the Polydrons and the one loose square.

Open the tetrahedron, move one of the triangles so that it looks like the illustration, and add the square, demonstrating how to snap Polydrons together. Fold it to form a square pyramid.

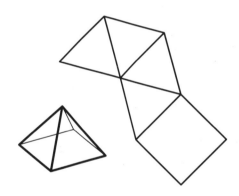

Exploration

Show a square and hexagon from the Polydron set. Ask what part of a polyhedron these represent. (Faces) Use the two pieces to again demonstrate how to snap the Polydrons together.

As students explore, they may construct larger polyhedra. After students have explored the construction set, let them to use the Polydron project booklet for more ideas on building complex polyhedra.

Polydron polyhedra are difficult to record. There are several options.

1. After completing a large polyhedron, have the student show it, identify it, and collect data to check it against Euler's equation. Then take it apart so that others may use the pieces for their constructions.

2. Use enlarged copies of polyhedra drawings that match students constructions for them to post. Most of these drawings can be found in the project booklet.

3. A three-dimensional computer graphics programs can be used to draw Polydron constructions. Printouts would provide records of solutions.

Closing Discussion

Again ask students to count faces, vertices, and edges of all polyhedra to test Euler's equation. This will be more difficult with larger polyhedra. To help them keep track, they can put a little piece of masking tape on each face, vertex, or edge. As they take the tape off, they can count the pieces.

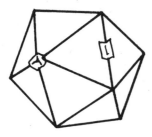

After the faces, edges, and vertices of all the polyhedra have been counted, ask whether there are any polyhedra that do not fit Euler's equation. If so, recount the faces, vertices, and edges for that polyhedron. Polydrons can be used to build some very irregular, concave polyhedra. Counting edges and vertices for these might be confusing because of reverse edges and vertices. Students are used to seeing edges and vertices as *peaks*, some edges and vertices of concave polyhedra are *valleys*.

To sort the polyhedra into regular, semiregular, and nonregular, have students count and identify the faces around each vertex.

Regular: On the icosahedron, five congruent triangles form each vertex, making it regular.

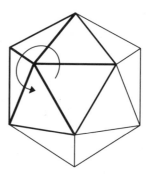

A regular polyhedron has the same number of regular polygons meeting at each vertex.

Semiregular: The truncated octahedron has two hexagons and one square around each vertex. Each vertex is the same, but there is more than one type of polygon at a vertex. The truncated octahedron is a semiregular polyhedron.

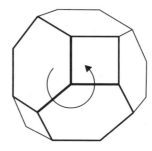

Nonregular: Except for the tetrahedron, all pyramids are nonregular polyhedra because the number of polygons that meet at a vertex varies.

For example, on the square pyramid, the vertex at the top is formed from four triangles, but at the base each vertex is formed from two triangles and a square.

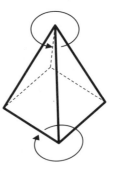

A truncated shape is one that has vertices cut off. With your students, look at an octahedron. Imagine a plane cutting off a vertex. If that plane is perpendicular to the axis of the octahedron, a square will remain after the plane has truncated the vertex. When all six vertices are cut off, the resulting shape, the truncated octahedron, has six square faces and eight hexagonal faces.

You or a student could use Polydrons to make a take-apart model of the truncation.

The truncated dodecahedron, the great rhombicuboctahedron, and the great rhombicosidodecahedron cannot be constructed with Polydrons because they require octagons or decagons. If you are using the teacher-made construction set, use the octagon and decagon to make these polyhedra.

Independent Investigation

Students may be interested in exploring and building geodesic domes. They can use commercially available geodesic dome kits or study a geodesic climbing dome on the playground, measuring enough of the edges to make a scale model out of straws. If two domes are constructed base to base, a geodesic sphere, a Bucky ball, is formed. Excellent illustrations of geodesic spheres of increasing size showing the arrangement of pentagons and hexagons are pictured in these articles.

Gary Taubes. "Great Balls of Carbon." *Discover* (September 1990): 52.

Robert F. Curl and Richard E. Smalley. "Fullerness." *Scientific American* (October 1991): 46.

Complex Polyhedra

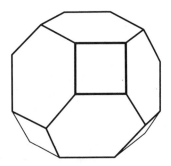

truncated octahedron

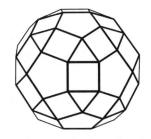

rhombicosidodecahedron

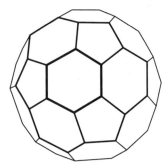

truncated icosahedron

snub dodecahedron

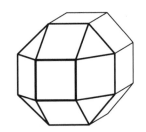

rhombicuboctahedron

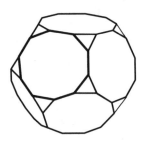

truncated dodecahedron

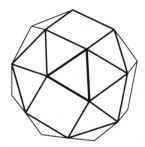

snub cuboctahedron

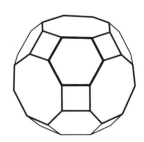

great rhombicuboctahedron

icosidodecahedron

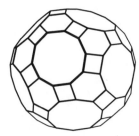

great rhombicosidodecahedron

Alternative Construction Set for Polydrons

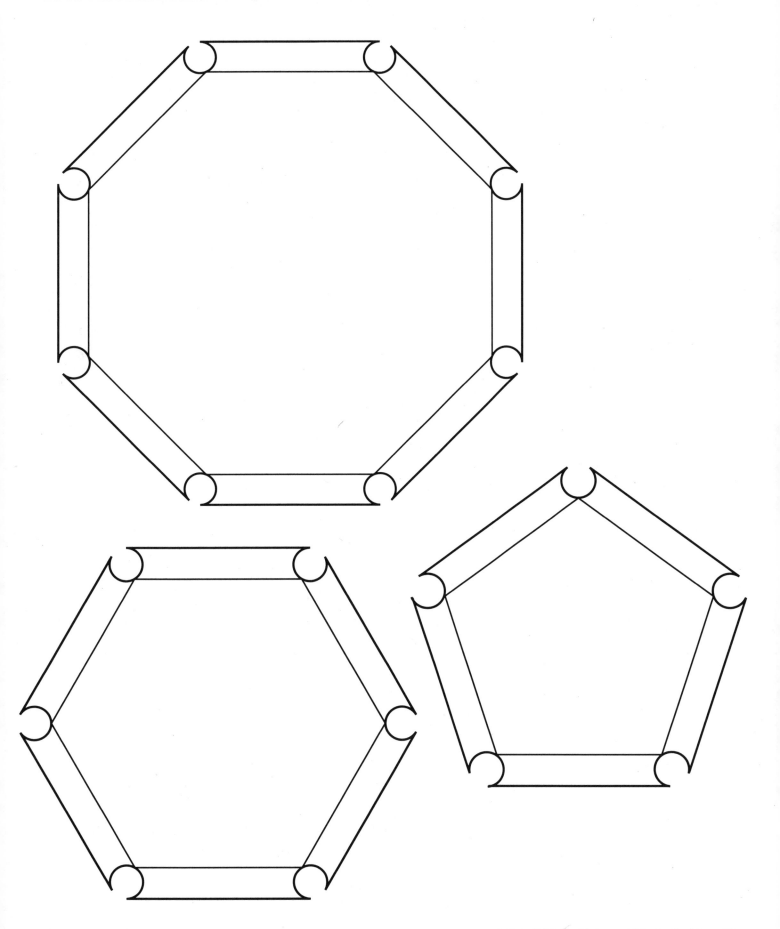

Alternative Construction Set for Polydrons

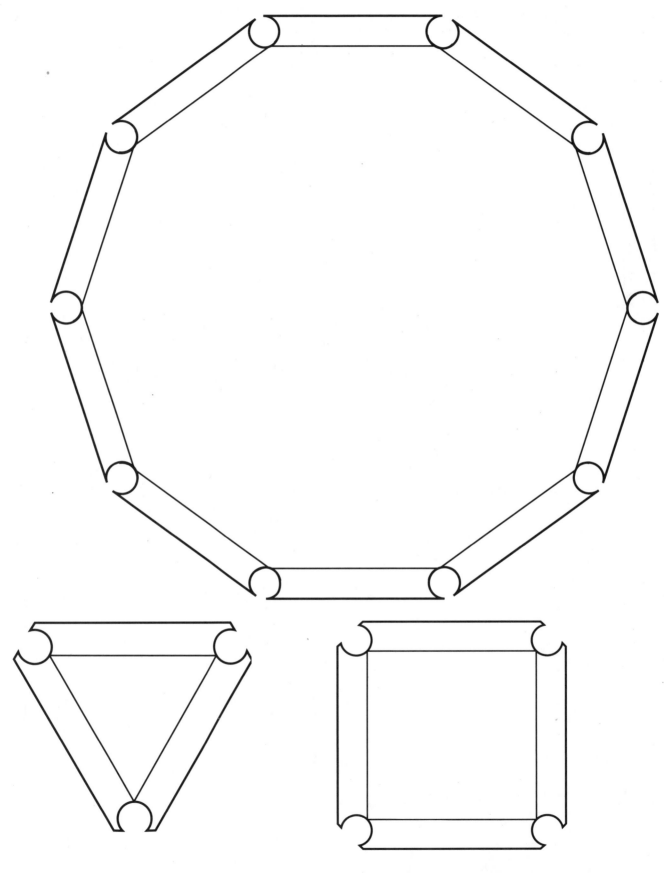

Exploration 5
Building and Exploring Tetrahedral Kites

Students will build and explore the characteristics of a variety of kites made from different complexes of tetrahedra.

Materials

Plastic soda straws
    Use rubber bands to bundle straws in groups of six.
Transparent tape
Copies of Kite Cover Pattern for Tetrahedron
    Page 35 is only half the pattern. Make two copies and tape them together at the dotted line.

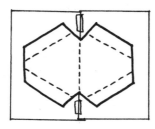

This cover fits a tetrahedron constructed from standard $7\frac{3}{4}$-inch straws. If you are using straws that are a different length, follow these steps to make a pattern.

1. *Use five straws to construct a rhombus with a diagonal.*

2. *Lay the rhombus on a large piece of paper and trace around it, leaving about a one-inch margin all the way around it.*

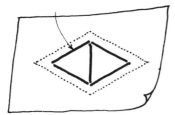

3. *Draw notches at the two vertices with the diagonal and truncate the other two vertices, forming four flaps. Do not cut this pattern out. Darken it in with a heavy pencil or black felt-tip pen.*

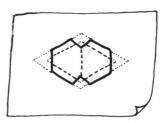

Wax paper covers, at least one for each students
    Cut about ten covers at a time by tearing off that many pieces of wax paper and stacking them on the cover pattern. Use a dull pencil pressing it while tracing. The wax will make the stack of paper stick together. Carefully lift up the whole stack and trim them all at once. Pull them apart and fold each one in half.
A model of a 4-celled tetrahedral kite
    See pages 26–27 for directions.
Kite string

Introduction

Tell your students about Dr. Alexander Graham Bell, who built tetrahedral kites in the early 1900s. Like several others, he was attempting to develop an airplane. To create an efficient lifting surface, Dr. Bell experimented with kite designs based on the tetrahedron and its multiple forms. He constructed a raft with layers of tetrahedra using wood and silk, with metal connectors and flew it from the top of a windy hill on a very long piece of rope. It required several men to handle the kite. Dr. Bell built the large kite at home, but he could take it apart and reassemble it at the launch site. When a kite crashed, he replaced the broken pieces from his repair kit, and continue his experiments. He found a lifting device that worked well, but when he attempted to add engines, the device would not fly. See his *National Geographic* article on page 85.

Show the kite you built in preparation for this session. These kites are made from straws instead of wood, wax paper instead of silk, and transparent tape instead of metal connectors; but they will also fly. Explain that there is a special way to tape the vertices so the kite doesn't pull apart in flight. Begin the exploration by building step by step with your students their first tetrahedra cells.

Exploration

Distribute transparent tape and a bundle of six straws to each student.

Students must follow the steps carefully. Paraphrase or read each set of directions as you demonstrate it.

1. Begin by taping two straws together. Use about one and a half inches of tape for each connection.

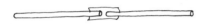

Leave a gap about the width of a straw. This prevents straws from being pinched or deformed when you bend the connection.

2. Add a third straw the same way. On the end of the third straw put a piece of tape as shown.

3. Bend the straws around, forming an equilateral triangle, crimping the two taped connections as you do to hold their position. Bend the straws so the taped connections are on the exterior of the triangle, as shown. Bend the last connecting piece of tape over, fastening the third straw.

4. Add another straw to one of the vertices. Apply tape as shown.

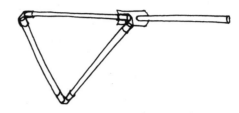

5. Add a second straw to the end of the first. Put another piece of tape on the end.

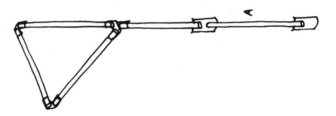

6. Bend the two straws around as shown to form two equilateral triangles, sharing the same base. Press the remaining piece of tape over, connecting the end of the fifth straw.

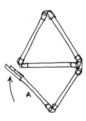

7. Add one last straw to one of the two free vertices. Put a piece of tape on the end of it.

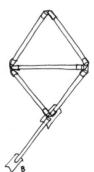

8. Bend the last straw up, crimping the tape into place. Now bend the other triangle up and press the tape over, making the last connection.

Distribute the wax paper covers and show how to cover two adjacent faces on the tetrahedron. Paraphrase or read these directions to students as you demonstrate each step with your tetrahedron.

1. Place the tetrahedron on the cover so that one straw is on the fold.

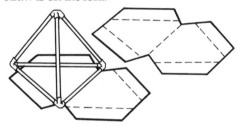

2. Fold the two flaps over the straws and tape them in place. Keep the cover tight, but don't distort the straws.

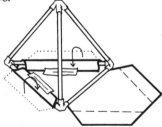

3. Flip the tetrahedron over onto the other half of the cover. Keeping the cover tight, tape the two remaining flaps in place.

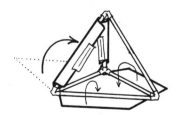

Individual Kites

If you have enough materials, each student can build a small kite composed of four tetrahedra. Distribute enough materials for each student to construct three additional cells independently.

When all students have completed their four covered tetrahedra, demonstrate how to assemble them into a larger tetrahedron, completing the kite.

1. Connect three tetrahedra with long pieces of tape. Make sure all the covered triangles are facing the same direction.

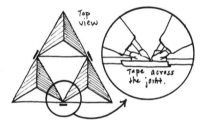

In this figure, cells A and B are facing the same way, but cell C is turned sideways.

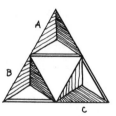

2. Place the last cell on top of the three bottom cells.

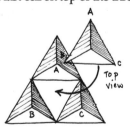

3. Tape the last cell to the tops of the three bottom cells.

Give each student a short length of string to test-fly their kites. Tie the string to the very top vertex of the kite.

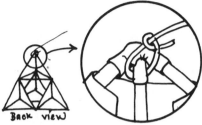

Demonstrate for your students how to fly a tetrahedral kite.

1. Place the kite down on the ground, with its front into the wind—the front of a cell is the edge where the two adjacent covered faces meet.

2. Walk away from the kite, unrolling about 15 to 20 feet of string.

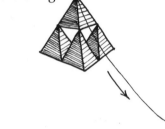

3. Gently pull the string taut and take a few running steps against the wind. At the same time, tug the string slightly.

   If the kite tips over on its side, don't drag it. Stop running and carefully set it up again.

4. Once the kite is in the air, if it begins to drop, pump the string gently, and it will begin to climb again.

Tetrahedral kites do not need tails. If the kite keeps dipping to one side you need to balance it by adding a little tape to the other side. The smaller the kite, the less stable it is.

Class Kites

From four small kites assemble a larger kite.

1. Tape three of the four smaller kites together just as you taped 3 cells together.

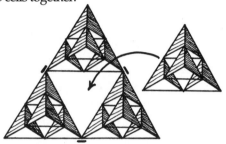

2. Place the fourth kite on top of the three kites and tape it into place.

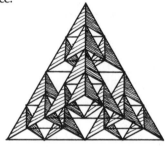

Show the kite and ask how many cells it takes to construct this kite. (16) Have students extrapolate to the next larger kite. Ask them how many cells it would take if we built three more of these and assembled a huge kite. (64) There is a photograph of a kite this size built by Alexander Graham Bell in his article on tetrahedral kites in the June 1903 issue of *National Geographic Magazine*, reprinted on pages 85–97.

In the article Dr. Bell states, "The most convenient place for the attachment of the flying cord is the extreme point of the bow.... A good place for high flights is a point halfway between the bow and the middle of the keel."

Tie the string to the top of the large kite and follow the flying instructions. Once the kite is up, give students turns handling the line.

After flying the kite with the string tied at the top, tie it at a point halfway between the "bow and the middle of the keel" and fly it again. For the 16-cell kite, tie it at the second cell down from the top.

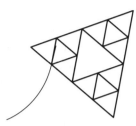

Ask how tying the string halfway between the top and the middle changes the way the kite flies. The students may notice that the kite flies higher and closer to a point overhead when the string is tied lower.

There are several ways to arrange tetrahedral cells to create kite forms. One of these is a solid arrangement. Introduce this solid arrangement of cells by first encouraging the students to look at the large open area inside the 16-cell kite. Ask what shape the large interior space is. It is an octahedron. If the students have trouble visualizing the shape of the interior space, draw an octahedron on the board, and trace the edges of the straws outlining the octahedron.

Explain that the next kite doesn't have large open spaces.

Solid Pyramids

1. Begin with a layer of tetrahedral cells.

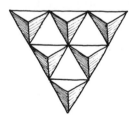

2. Add a second layer by taping one cell on top of each triangular group of cells below it.

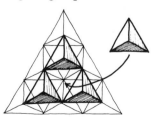

3. Add the last cell to the top three cells.

Ask, how many tetrahedral cells there are in this kite. (10) If a fourth layer is added to the bottom or base of the kite, how many more cells would be needed? (10) Analyze the kite's structure to help students find the answer.

Draw the first three layers on the board.

Ask what the pattern is in the number series: 0, 1, 3, 6. (Add one, add two, add three, add four; add one more to the amount you add each time.)

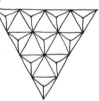

$$0 + 1 = 1$$
$$1 + 2 = 3$$
$$3 + 3 = 6$$
$$6 + 4 = x$$

Ask what the next number in the series would be. (10) Test the answer by building or drawing the fourth layer. How many cells would a four-layered kite have? (20)

The fifth layer would contain 15 cells. A five-layered kite would have 35 cells.

Use the number series to predict the number of cells in other layers. For example, the tenth layer would have 55 cells.

Fly the three- or four-layered class kite. If your kite crashes, just remove the bent straws and replace them. If a cover is not badly torn, mend it with transparent tape. Reinforce the top vertex of the cell where the string is tied.

Compare the flight of this kite with the string tied high (point A) and lower (point B).

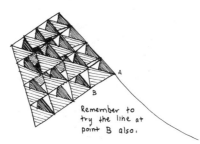

Remember to try the line at point B also.

Rafts

A raft is a rectangular arrangement of tetrahedral cells. It has the disadvantage of the extra weight from the bracing, but it still flies well.

1. Begin with three cells in a row, covered sides down. Tape them together as shown.

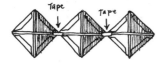

Tape   Tape

2. Tape another row of three and carefully turn both rows over, taping them together as shown.

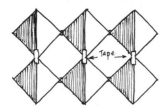

Tape

3. Repeat steps one and two, then tape both sets of two rows together. There should be four rows of three cells taped together.

4. Then add five pairs of bracing straws as shown.

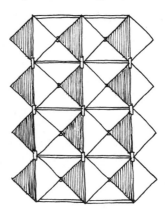

5. Carefully turn the raft over, adding four triples of bracing straws in the other direction.

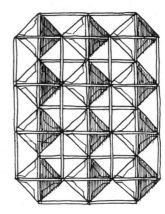

To fly a raft kite, two pieces of string are needed. Cut a piece of string (A) and tie it as shown so that, when pulled taut, it will form an equilateral triangle with the top of the kite. Attach the kite string (B) to it as shown.

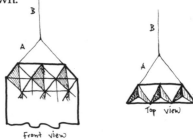

Front view   Top view

This completes the one-layer raft, launch it in the position shown.

Launch position

side view

The only bracing needed in multilayered rafts, is on the top as shown in step 4, and the bottom as shown in step 5. In layers between, the cells' edges replace the braces.

If you add a second layer to the top of the three by four raft, use a two-by-five array.

The raft is actually a slice from a tetrahedron.

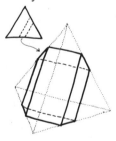

Single rows or files can be flown by themselves.

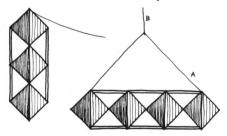

A single cell will only spin when flown alone. You need at least 2 cells for stability.

Modified Raft

The cells may be arranged in alternating rows as shown.

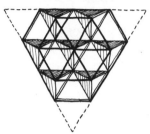

This forms a single layer which is a truncated slice from the base of the tetrahedron.

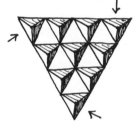

If you didn't truncate the corners, you would have a complete triangular slice.

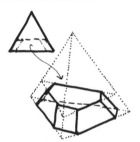

To attach string, tie the first piece of string at the kite's widest point.

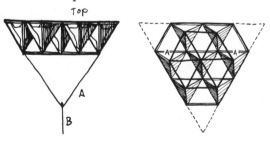

Explore a variety of tetrahedral combinations. Test them to see which of them fly well.

Hexagonal Box Kites

Unlike kites built of single covered tetrahedra constructed by individuals and combined to become part of a class kite, this tetrahedral, hexagonal box kite is one structural unit, built layer upon layer. Once you have started the construction, pairs of students can take turns adding to the framework. Or you can build the kite and show it as an example of a tetrahedral application for Bell's triangular-celled box kites.

Share this information from Bell's article with your students: Lawrence Hargrave, an Australian, invented the box kite in 1893. The kite was made of two right rectangular prisms separated by a large space.

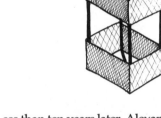

Less than ten years later, Alexander Graham Bell improved the Hargrave box kite, applying Hargrave's principles to triangular structures and later to tetrahedral forms.

Either show the completed hexagonal box kite or show how to complete the first section by doing steps 4 through 7 with the class. (Finishing steps 1–3 before will save class time.)

How to Build a Hexagonal Box Kite

1. Construct a hexagon with its diagonals. If it doesn't lie flat, gently pull at the vertices to lessen the tension.

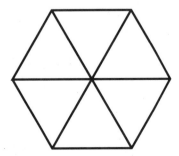

2. Build a tetrahedron on three alternating triangles.

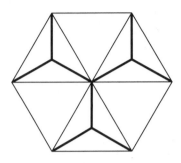

3. Connect the top vertices of the tetrahedra to form a triangle. This forms the smallest modified raft.

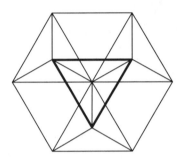

4. Add two straws to each of the two edges of the triangle as shown, forming two additional triangles. You will have half of the next hexagon.

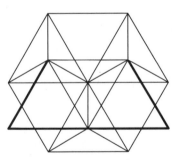

5. Add two straws to the free vertex of each new triangle, forming two octahedra.

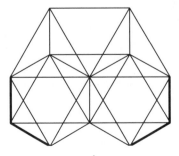

6. Complete the second hexagon by adding five straws as shown.

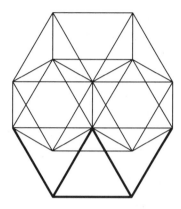

7. Add two straws, forming two tetrahedra and completing this layer.

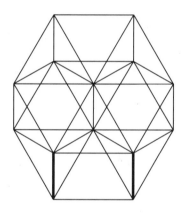

Now go back to step 1 and build a new layer. Continue building until you have completed four layers.

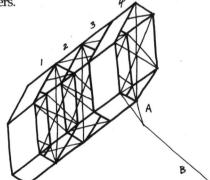

Discuss with your students the geometric name for the shape of this framework. (*Oblique hexagonal prism*; see page 16.) Bell's early triangular box kites were *right hexagonal prisms.*

Prepare covers for the first and fourth layers. Tape two tetrahedral wax paper covers together. Make four pairs, two for the top layer and two for the bottom. (See page 36 for the tetrahedral cover pattern.) Tape on the four pairs of tetrahedron covers first. Then complete each hexagonal ring by taping on the four $7\frac{3}{4}$" squares. The side flaps of each square will overlap pairs of tetrahedral covers already in place.

To fly the kite attach the string to the top of the top layer. Note the launch position.

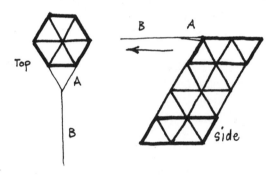

The class may want to cover additional interior faces making the tetrahedral form look more like Bell's original triangular-celled box kite. They can include the performance of this modified hexagonal box kite on the kite-performance chart.

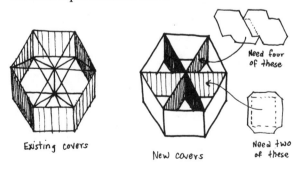

Existing covers   New covers   Need four of these   Need two of these

Closing Discussion

Students have had opportunity to build and fly a variety of tetrahedral kites. In the process they have discussed some results. For example, they analyzed the effect of line attachment position on kite performance. Now they can compare the kites built and flown summarizing all the data on a chart. Ask students to suggest ways to compare kites. Write comparison criteria across the top of a chart and list types of kites down the side.

A stable kite is one that usually remains level in flight with little dipping or rotation. A kite with good lift will fly easily even in light winds.

When making lists of kite types, include several kites in the same category but provide room for each on the chart. For example, under solid pyramids, begin with the simple 4-celled form and include larger complexes. Rafts could include lines and files, or modified rafts.

Use students' subjective evaluations to complete categories like stability and lift. Students may use terms such as excellent, good, fair, or bad. Use words that reflect a consensus of opinions.

When the chart is completed, look for any patterns in the data. For example, students may notice that the fewer cells a kite has, the less stable it is. Or the greater number of extra straws a kite has for bracing, the less lift it exhibits since bracing straws add extra weight. Others may notice that pyramids with an open space and box kites were the most stable in flight; both have lifting surfaces separated by large spaces.

Try conducting a popularity poll to determine the most popular kite form. Students may vote for the pyramid with the large octahedral space because the symmetry makes it aesthetically pleasing, or for the hexagonal box kite because it is unique. These considerations may play as big a part as stability or lift in the decision.

Independent Investigation

Students may experiment with kite forms at home. Have them bring their variations and extensions of forms to school.

Type of Kite	Number of Cells	Number of Braces	Stability	Lift
Pyramids with an Open Space				
Solid Pyramids				
Rafts				
Hexagonal Box				

Students may design variations of box kites just as A. G. Bell built variations of the Hargrave box kite.

Students may want to show commercial kites such as Tetra Kite.

Research on kites could include reading about and reporting on the old weather kites. Scientists used kite trains to raise weather kites to great heights. Encourage student experimentation such as flying a train of several box kites, one line tied onto the next kite, a kite flying a kite.

A student could take a straw model of one of the solids on page 16 and attempt to make a kite out of it. By trial and error, the students could determine the best faces to cover by experimentally covering various combinations of faces with wax paper. They can also experiment to find the best place to attach the string.

Kite Cover Pattern for Tetrahedron

This is only half of the pattern. Make two copies and tape them together at the dotted edge.

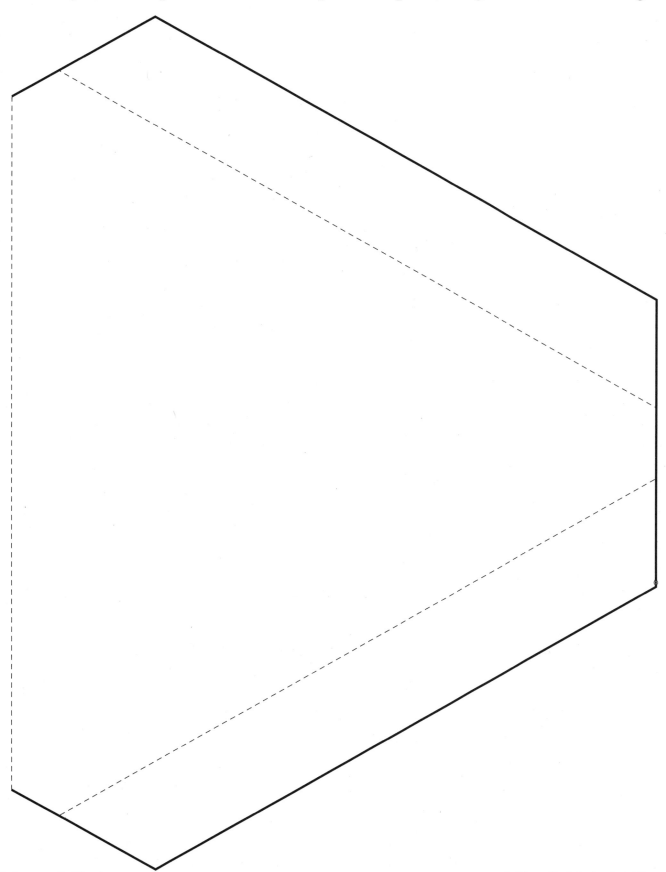

Cover Pattern for Hexagonal Box Kite

This is only half of the pattern. Make two copies and tape them together at the dotted edge.

Exploration 6
Finding the Height of a Kite in Flight

Students will use a simple, homemade sextant to measure the angle of a kite in flight and compute its height, geometrically or trigonometrically.

Materials
Homemade sextants

Make, or make with your class, one or more sextants. Each sextant requires a protractor, a straw, string, a weight, and tape. To make a sextant, tape a straw to the bottom of a protractor and tie a string through the hole. Tie a weight to the string.

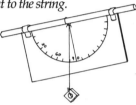

Copies of Protractor Pattern (page 39)

Copy the pattern onto card stock or glue paper copies to tagboard. Pre-punch holes for the string.

Soda straws

7" pieces of string

Weights, such as nuts or washers

Transparent tape

Introduction
Explain that a sextant is a tool used to find the angle between two distant objects, the ground and the object whose height or altitude you want to measure. Students will use their sextant to find the angle between the ground and a kite in flight. Using that flight angle, they can calculate the kite's altitude.

Exploration
Facilitate as students, individually or in groups, make their sextants.

Demonstrate how to use the sextant. Without poking your eye, look through the straw, aiming at the object, and gently press the string against the protractor. Carefully turn the sextant and read the number of degrees marked by the string. Subtract that number from 90° to find the flight angle, the angle between the ground and the kite string. In the illustration, 90° − 80° = 10°.

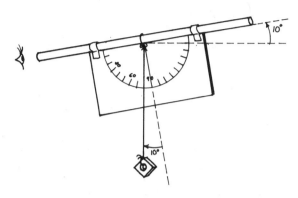

Find a tree or building that can be seen from the playground for all students to measure. Assist them to accurately sight the object through the straw, mark the angle, read the angle from the protractor, and calculate the angle between the ground and the top of the object. Discuss the results and the reasons for any differences between measurements.

Students can now fly their kites and measure the flight angle for each kite by using the sextant to sight along the string.

To find the altitude of a kite geometrically, students need to measure two other things.

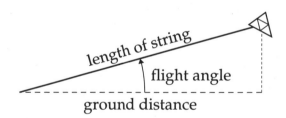

1. Length of the string. Use a new ball of kite string so you'll know the length.

2. Ground distance to the kite. Mark the place where the student flying the kite is standing. Have another student walk out until he or she is directly under the kite, marking that place. Measure the ground distance between the two points.

As students gather and record data, they can organize it in a table.

Kite Form	Flight Angle	Length of String	Ground Distance

To find the altitude of a kite using trigonometry, students need the flight angle and either the ground distance between the student flying the kite and a student standing under the kite itself or the length of the string.

Students can gather information on several kites, they can use one kite with the string attached at the highest point and then with it tied between the highest point and the middle, or they can use one kite flown at different times.

Closing Discussion

Help students compute the altitude of each kite.

To compute the altitude geometrically, make a scale drawing. For a demonstration at the board, use the scale one inch equals ten feet. Draw a line. Mark point A, measure the ground distance, and mark point B. Measure angle A, the flight angle. Draw $\overline{AC}$, measure the length of the kite string, and mark point C. If $\overline{BC}$ is not perpendicular to $\overline{AB}$, the students' measurements are inaccurate.

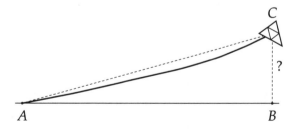

If $\overline{BC}$ is approximately perpendicular to $\overline{AB}$, measure $\overline{BC}$ and use the scale to find the altitude of the kite. Students may point out that point A is not actually on the ground because someone is holding

the string. Three or four feet can be added to the altitude to account for the person holding the string above the ground.

Students can apply the Pythagorean theorem to find the altitude of the kite without using the flight angle.

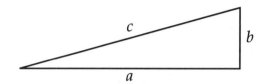

$$a^2 + b^2 = c^2 \ \text{ or } \ b^2 = c^2 - a^2$$

If you are computing the height trigonometrically and have measured the ground distance, use a trigonometric table or calculator to find the tangent of angle A and solve for BC.

$$\tan \angle A = \frac{\overline{BC}}{\overline{AB}}$$

If you are computing the height trigonometrically and have measured the string length, find the sine of angle A, and solve the equation for BC.

$$\sin \angle A = \frac{\overline{BC}}{\overline{AC}}$$

As you explore the results with the class, talk about the accuracy of their measurements. If the location of the string attachment made a difference in a kite's altitude, compare that difference to other kites to determine the best placement for the string. Ask which kite form showed the greatest change in altitude when the position of the string was changed.

Career Awareness

Invite a surveyor, perhaps from the State Department of Highways, to visit your class, demonstrate a transit, and talk about what surveyors do.

Independent Investigation

Students may use their sextants to find heights of trees, buildings, or other landmarks.

Protractor Patterns for Sextants

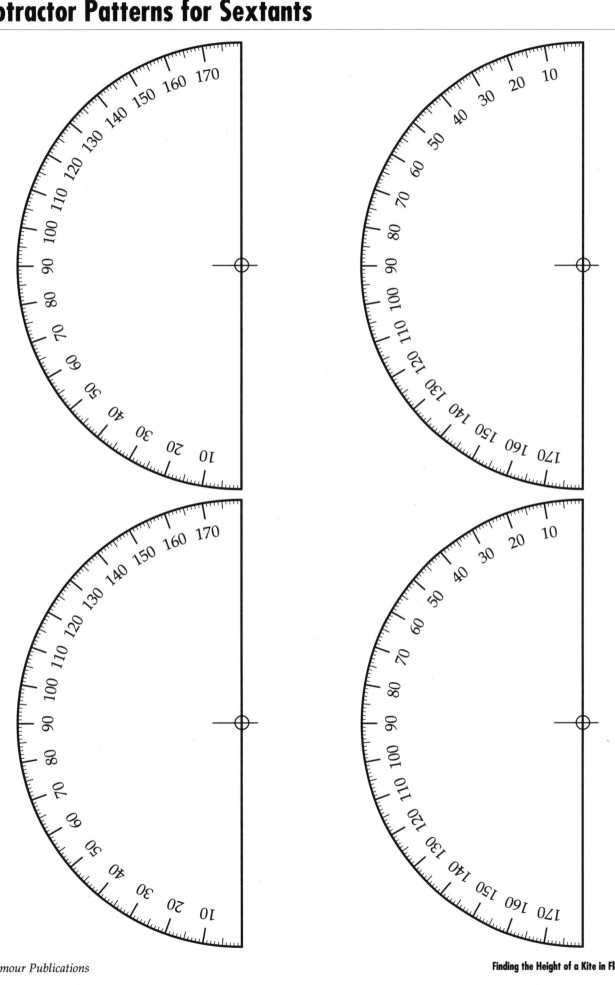

Exploration 7
Building Grid-Paper Boxes

Students will use centimeter grid paper to build models of prisms, use manipulatives to find volume of prisms, develop equations for finding volumes of prisms, and compute volumes of prisms.

Materials
Scissors
Transparent tape
Centimeter grid paper
   Copy page 43 on light-colored card stock.
Rectangular grid paper
   Copy page 44 on light-colored card stock.
Centimeter Cubes
Black construction paper, 9" × 6"
Soda straw models of cubes and rectangular prisms
   from Exploration 2
Demonstration model of a 5 × 4 × 3 rectangular
   prism
   Leave the top loose as a lid.

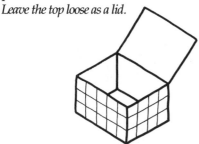

Demonstration model of a triangular prism with a
   base that is one half of a 4 × 4 square with a
   height of 5
   *Use centimeter grid paper for four of the faces and the
   rectangular grid paper (page 44) for the slanted face
   (largest rectangular face).*

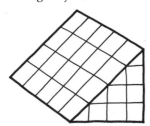

Six pieces of 6 × 6 grid paper for a demonstration

Introduction
Show straw models of cubes and rectangular prisms. Explain that students will build models of prisms using centimeter grid paper. Show the example 5 × 4 × 3 model made of grid paper.

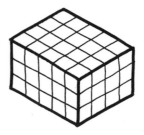

They will use centimeter cubes to find the volume of these prisms.

Exploration
Distribute transparent tape and a sheet of centimeter grid paper to each student. Have them cut six 6 × 6 squares from the edges of the grid paper. Demonstrate how to tape them into a prism called a *cube,* or *regular hexahedron.*

1. Tape four squares in a row.

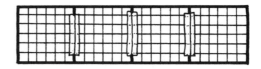

2. Add a square to each side, forming a cross.

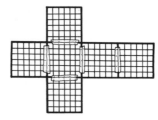

3. Turn the cross over and fold the bottom square up and the top square down until they meet, then tape them.

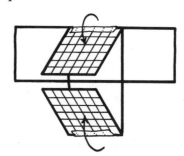

4. Tape one of the other two squares to form a 6×6×6 cube with a lid.

With the remaining grid paper, students can build other prisms. Most students will build cubes. Have them try other rectangular prisms.

Open the lid of the 5×4×3 prism and place a few cubes inside. Ask students to predict the number of cubes it will take to fill the prism. (60) Fill the prism with cubes, then dump them out and ask students to count them. Compare the result to the predictions. If someone knows how to find the volume by multiplying length times width times height, ask them to not explain that yet, they will have an opportunity later. It is important that other students discover the concept by exploring with the cubes.

Explain that they will build other boxes, fill them with cubes to find their volumes, and display the prisms with their dimensions and volume. Use your demonstration model to show students how the prisms will be displayed. Tape down the lid and tape or glue the 5×4×3 demonstration prism to a piece of 9"×6" black construction paper. Using a white crayon, write the dimensions of the prism and its volume on the paper.

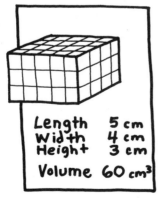

Length   5 cm
Width   4 cm
Height   3 cm

Volume   60 cm³

Distribute additional pieces of grid paper so students can build prisms of varying dimensions and find their volumes.

As students complete prisms and mount them with the dimensions and volume, post them on a large bulletin board.

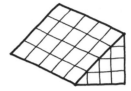

During one of the work periods, show your class the triangular prism you constructed. Point out how the grid lines match along the edges, even on the sloping face, because you used the rectangular grid paper (page 44) for that face.

Have students build triangular prisms by cutting a square in half for the two triangular bases, cutting rectangles of centimeter grid paper for two of the faces, and cutting a rectangle of the rectangular paper for the sloping face. Some students will try to fill their triangular prisms with centimeter cubes to find the volume. Allow them to explore and try to work out a way of finding the volume.

Closing Discussion

When all students have completed and posted at least one prism, evaluate with the class the posted data.

Ask whether all the lengths, widths, and heights are counted correctly. Ask how they can tell whether the volume is correct. A student may suggest finding duplicate prisms and comparing their volumes.

By now someone will notice that you can find the volume by multiplying length times width times height. Discuss why this equation works. You may need to introduce the concept. The posted volumes can be checked using the equation $l \times w \times h = V$.

Ask whether it is possible to have two prisms with the same volume, but different dimensions. (Yes) If no examples of this are on the bulletin board, refer to your 5×4×3 sample prism. List other factor triplets that are equivalent to sixty, for example: $2 \times 2 \times 15, 3 \times 2 \times 10, 3 \times 1 \times 20,$ or $2 \times 6 \times 5$. In the process, students may discover other prisms they want to construct and add to the collection.

With your students, look at the triangular prisms. Ask, how students could find the volume of a triangular prism. Students may have already reasoned that it would be one half of the volume of a prism with the same dimensions. Put two triangular prisms of the same dimension together so that all the students can see that the volume of one is one-half the volume of the corresponding square prism.

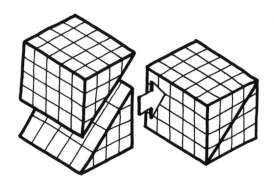

The equation is $V = \dfrac{l \times w \times h}{2}$.

This equation can be applied to all right-triangular right prisms, but will not work for all triangular prisms.

Using a diagram like this drawn on the board, review how to find the area of a triangle.

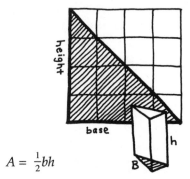

$A = \frac{1}{2}bh$

Explain that the area of the triangle is the base of the prism. To find the volume of any right prism, find the area of one of its bases (B) and multiply it by the height of the prism (h), yielding the equation $V = Bh$.

Independent Investigation

Students enjoy sorting the posted prisms into groups, such as cubes, triangular prisms, and rectangular prisms. These groups can be organized by size, and students might realize that some cubes are missing. Looking at the series of cubic numbers, they could predict the volumes of missing cubes.

As a class, students listed the possible dimensions for prisms with volumes of 60 cm³; individuals can list and construct rectangular prisms with volumes of 144 cm³. Set aside a special bulletin board for the many prisms with the volume of 144 cm³. Students can add to this display after the class has moved on to other explorations.

Students may be interested in finding the volumes of prisms having bases that are parallelograms, hexagons, or octagons.

Centimeter Grid Paper

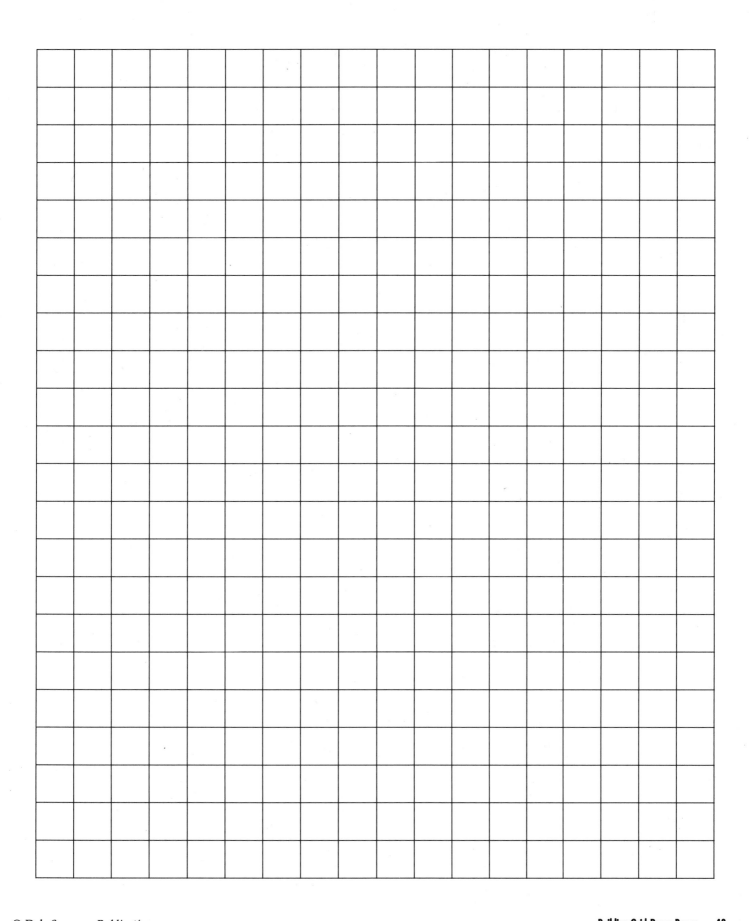

Rectangular Grid Paper

Exploration 8
Comparing a Cube and Its Related Tetrahedron

Students will explore the relationship between a cube and the tetrahedron whose edges are the diagonals of the square face of the cube. This activity, involving pouring and measuring sand, works best as a small group activity or as a demonstration.

Materials
Cardboard models of a cube and its related tetrahedron

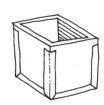

Use shoe box cardboard, masking tape, and white glue to build sturdy models of a 4" cube and its related tetrahedron (patterns on page 47). Leave one side open so models can be used as containers for sand.

Half-gallon of clean sand

Diagram or straw model of the tetrahedron in the cube

To build the model, construct the 4" cube first, then add the diagonals that are the edges of the tetrahedron. Using a different color straw for the diagonals will highlight the tetrahedron.

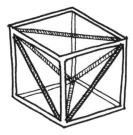

Construct the diagonals by cutting three straws in half, cutting slits about one and a half inches long in one end, and fitting each one inside a whole straw.

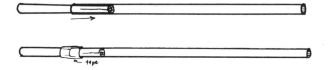

Place the straw cube inside a shallow cardboard box and square the base by pushing it into a corner. Tape one end of the lengthened straw to a vertex of the base.

Slide the slit piece of straw in or out, making it equal the diagonal length of the square base. Tape the sliding joint to fix its length, and tape the free end to the opposite vertex of the base. Use this diagonal as a standard and fix the length of the other diagonals. Then tape the remaining diagonals in place to complete the model.

The five solids that combine to form a cube
Make the tetrahedron and four triangular pyramids from the patterns on pages 48–49. Or you could use five of the pyramids from the Polyhedra Blocks Basic Kit.

Introduction
Show the soda straw model illustrating the relationship between a cube and the tetrahedron formed by the diagonals of its faces, or show a two-dimensional diagram on the board or overhead transparency. Two transparencies, one with the cube and one with the related tetrahedron, could be shown separately first and then combined to show how the tetrahedron fits into the cube. For most students the three dimensional model will be easiest to understand, but showing the relationship more than one way will increase understanding.

Explain that a tetrahedron can be made from diagonals of the six faces of a cube. The six diagonals of the faces of a cube form a tetrahedron.

Use the straw model to point out that each edge of the tetrahedron is a diagonal on a face of the cube.

Show the model of the five solids that can be assembled into a cube. Take the cube apart and remove the tetrahedron. Point out the four leftover nonregular pyramids.

Ask students to think about how much greater the volume of the cube is than the volume of its related tetrahedron. (Three times, but don't tell the students yet; allow them to discover it.)

Explore with the students how to compare the volumes of the cube and its related tetrahedron. Show the two containers you made, explaining that it is possible to compare the volumes by filling both cardboard containers with sand and comparing the two amounts.

Exploration

Demonstrate or use student volunteers to pour, measure, and record the amount of the sand each solid will hold. At an activity center, a few students at a time may go and pour sand, measure it, and record the results.

Closing Discussion

Ask students to look at the records of the volume of the tetrahedron and the square and find a relationship between them. Have them divide the volume of the tetrahedron into the volume of the related cube and discover that the volume of the tetrahedron is one-third that of the related cube.

Independent Investigation

Another polyhedron related to the cube is the cuboctahedron. Applying what was learned about the tetrahedron and the cube, students can investigate a way to find the volume of the cuboctahedron. As with the previous activity an open-sided cardboard pyramid equivalent to the truncated corner could be built and filled with sand. If the sand from eight of these were measured and subtracted from the sand that it took to fill the 4″ cube, you could find the volume of the related cuboctahedron.

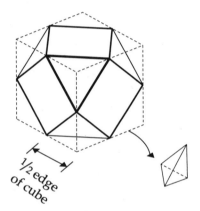

1/2 edge of cube

Patterns for Cube and Tetrahedron

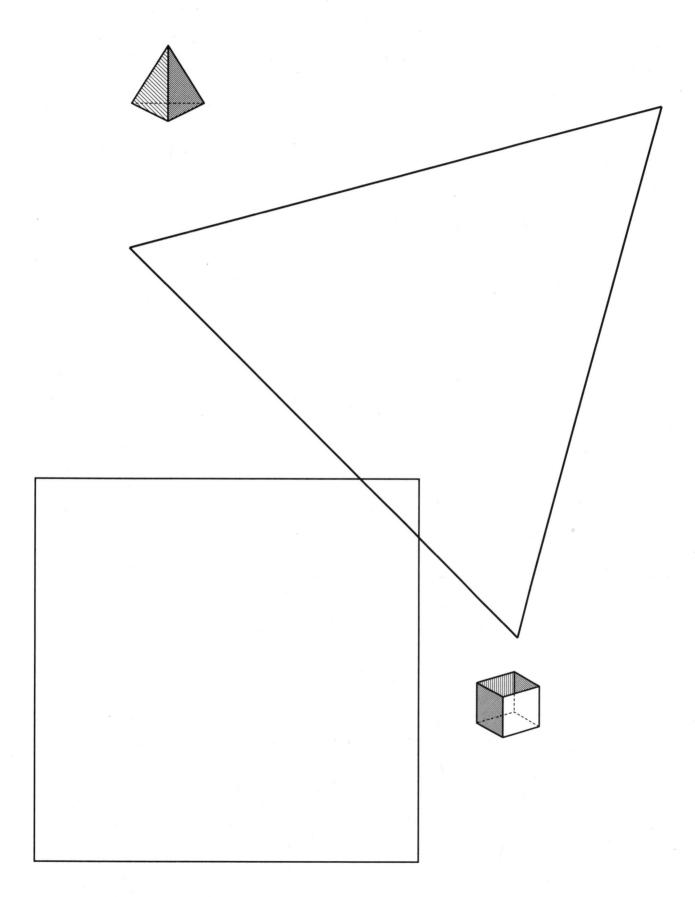

Pattern for Tetrahedron

Make two copies; reverse one and glue it to the other as shown in the illustration to make one tetrahedron.

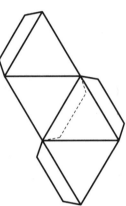

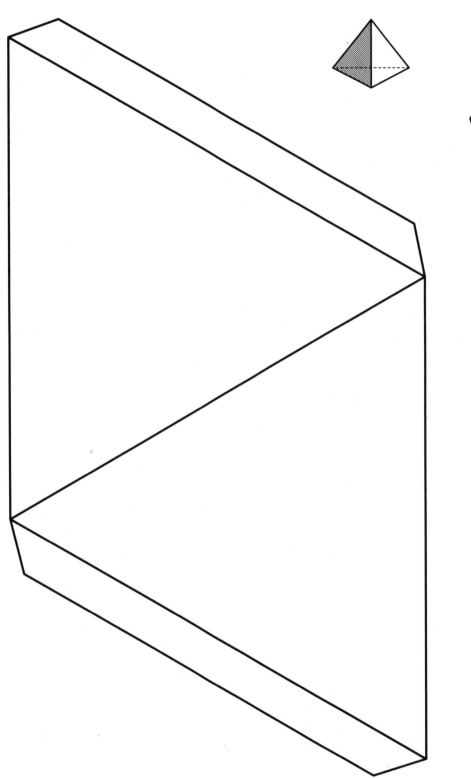

Pattern for Pyramids

Make four of these pyramids.

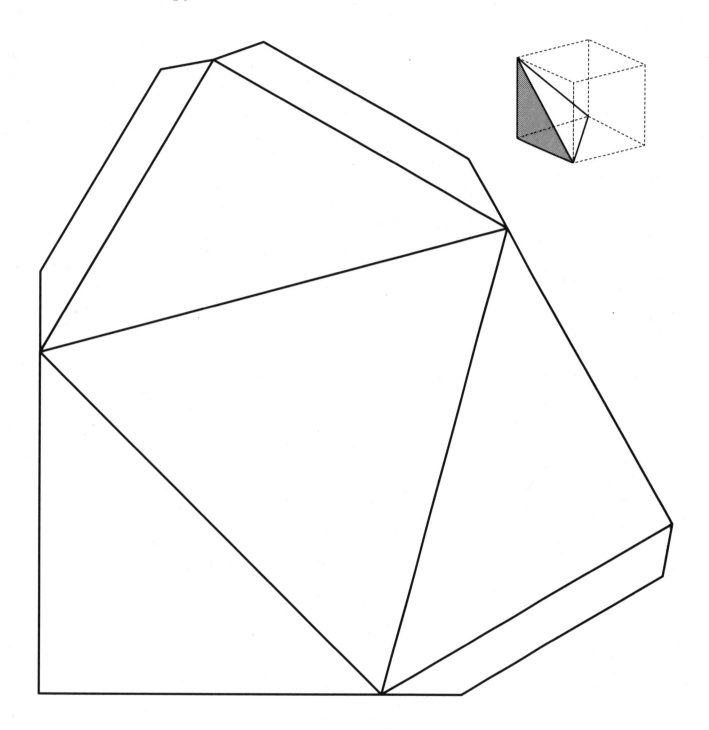

Exploration 9
Building Models of Pyramids

Students will build models of pyramids with grid paper and develop an equation for finding volumes of pyramids.

Materials
Centimeter grid paper (page 43), slope paper
  (page 44), and triangular grid paper (page 52)
Copy on light-colored card stock.
Cardboard tetrahedral container from
  Exploration 8
Cardboard container of tetrahedron's
  related prism
  *Make with shoe box cardboard, masking tape, and
  white glue using the patterns on page 53.*
Two paper pyramid models from patterns
  *Use the patterns on pages 54–55 to make models
  with enlarged grid paper that students can easily
  see.*

Introduction
Show the two models of pyramids you
constructed. Show the type of slope grid paper
that was used to make each one.

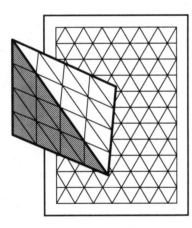

Explain that the base and all perpendicular faces
are made of centimeter grid paper. It may help to
show the original patterns on pages 54–55 to
show how each one looks flattened out.

Explain that they will use several types of grid
paper to build pyramids.

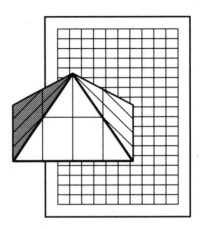

Exploration
Distribute materials allowing students to explore
and construct several pyramids. As they draw
diagonals on the grid paper, have students use a
straight edge. Remind the class to cut on the lines
or diagonals, otherwise it will be difficult to find
the area of a pyramid's base or its height. The
lines on various grid papers must always match
along the edges of the pyramid. All bases and
perpendicular faces should be cut from the
centimeter grid paper. Careful cutting and taping
is necessary for neat and accurate models.

Closing Discussion
Have students share some of their pyramids.
Evaluate each pyramid by asking whether the
base and perpendicular face are made of
centimeter grid paper, and whether all cuts have
been made on the lines or straight diagonals.
Check to see whether the lines match up on the
various grid papers. Using the chart of pyramids
on page 16, have students identify the pyramids
they have constructed. If several students have
constructed the same type of pyramid but in a
different size, point out that they are called *similar
pyramids*. Ask the students to save the pyramids
for a later activity or have them write their names
on the pyramids and collect them for use later.

Review results of the previous experiment with
the cube and the related tetrahedron. The volume
of that tetrahedron is one-third of the cube's
volume.

Use sand to demonstrate that it takes three tetrahedral containers to fill a right-triangular prism container with the same base and height as the tetrahedron. Students can explore this independently before you show it to the class.

Have students recall the equation for finding the volume of a prism, $V = Bh$.

Write these steps in on the board so students can see the reasoning.

$V = Bh$ is the equation for finding the volume of a prism.

Ask students to express the relationship between the volume of the tetrahedron and its related prism. Write their statements on the board. Sample responses:

> "Three times the volume of a tetrahedron equals the volume of the related prism"

> "The volumes of three tetrahedra equal the volume of the related prism."

Ask what the volume of one tetrahedron equals. Write their responses on the board. Sample response:

> "The volume of the tetrahedron equals one-third the volume of the related prism."

Cross out the phrase "the volume of the related prism" and substitute Bh in its place and finally shorten it to

$$V = \frac{1}{3}Bh.$$

Have students use the equation to compute volumes of pyramids they constructed. Use the grid lines for measurement.

Have other students compare the computations for accuracy.

Students who constructed similar pyramids may want to compare their volumes. How would doubling the measurements affect the volume of the resulting similar pyramid? (If each measurement is doubled, the volume is multiplied by eight.)

Independent Investigation

Students may explore irregular pyramids. If grid paper is used, tape the pieces so that lines are inside; students will not have to worry about lines matching along the edges. If students cut the bases out of grid paper they can find the area of the base by counting the squares.

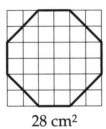

28 cm²

The octagon's area equals 28 cm².

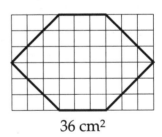

36 cm²

The hexagon has an area of 36 cm².

Students may build related prisms for their pyramids, filling them with sand to compare volumes.

Triangular Grid Paper

Pattern for Prism

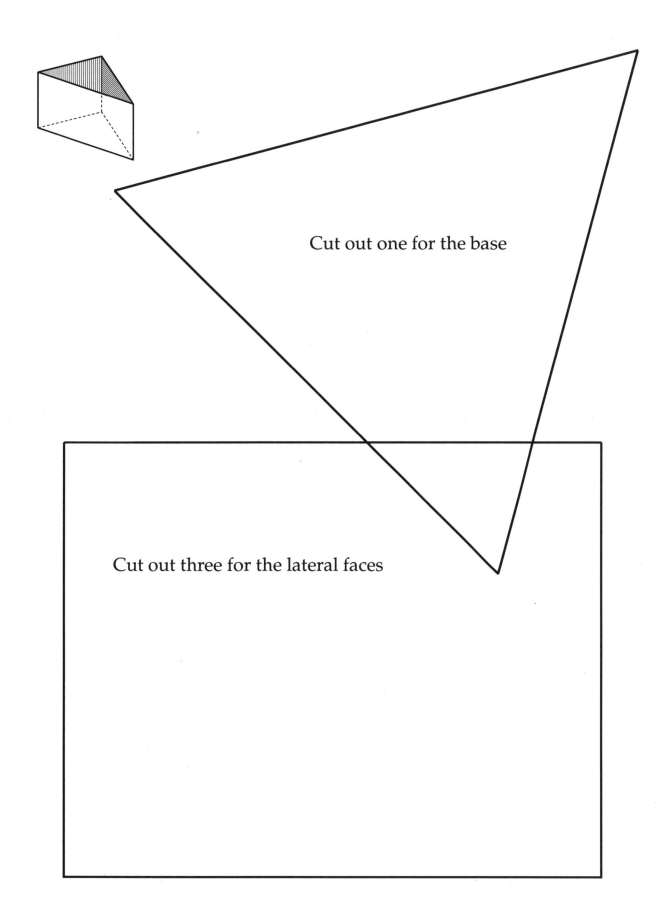

Cut out one for the base

Cut out three for the lateral faces

Pyramid Using Triangular Grid

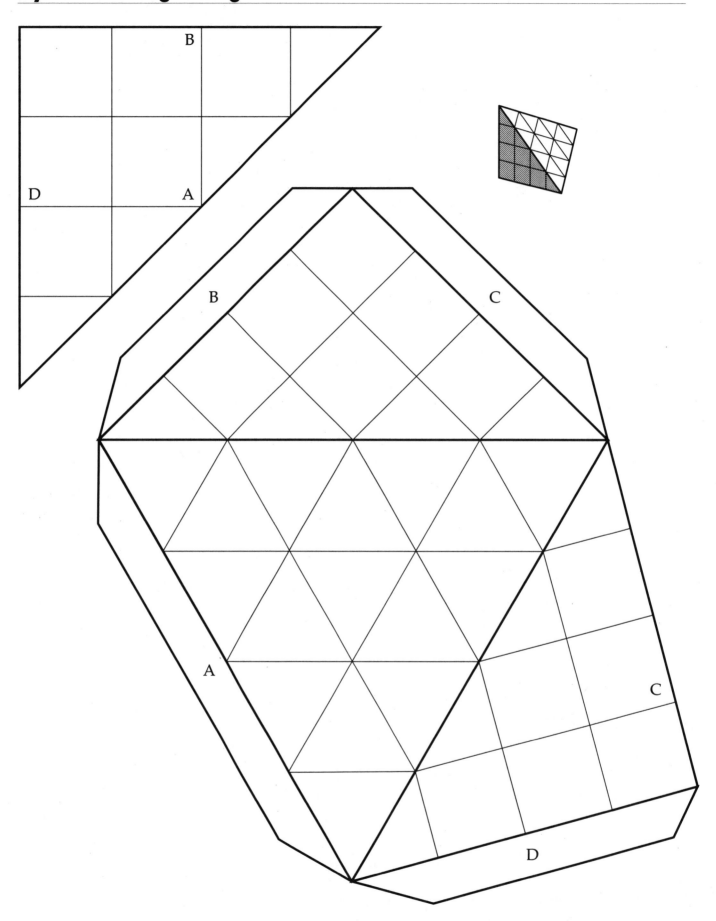

Pyramid Using Rectangular Grid

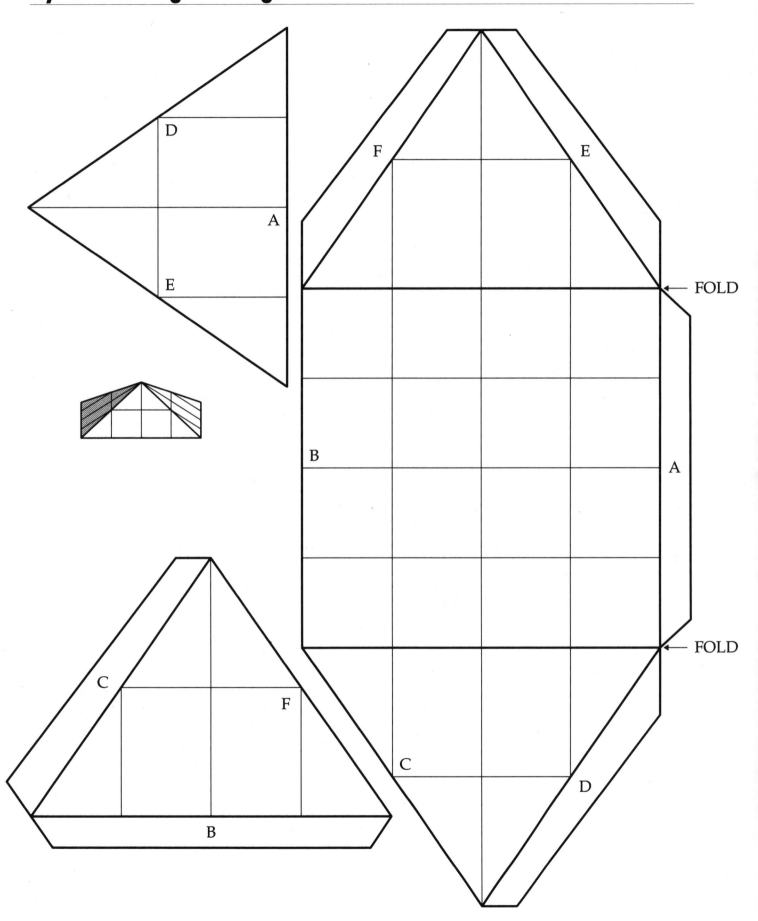

Exploration 10

Exploring Combinations of Pyramids

Students will combine pyramids to construct solids including similar pyramids, right and oblique prisms, and truncated solids. They will make paper models of several solids and compare them.

Materials

Polyhedra Blocks Basic Kit
> *Magnetic blocks in three sizes (A, B, and C) available from Dale Seymour Publications.* **Caution** *young children to handle plastic Polyhedra Blocks carefully, they have sharp points.*

Or teacher-made models of the same blocks
> Patterns on pages 58–59
> Card stock or other heavy paper
> Scissors
> Transparent tape or glue
> Double-sided transparent tape
>> *Double-sided tape on faces of the card-stock models serves the same purpose as the magnets on the plastic set.*

Introduction

Show several pyramids, demonstrating how to match up congruent faces forming larger solids. Explain that this set of blocks will be used to explore a series of problems. If this is the first time your class is using Polyhedra Blocks, allow students to explore the blocks informally.

Comparing Numbers of Faces

Materials

Polyhedra Blocks
Press-on dots
> *Use a color that can be easily seen.*

Heavy paper
Rulers
Scissors

Introduction

Display the blocks. Show block A, B, or C and ask the class how many faces it has. (Four)

Ask students to suggest a combination of two blocks that form a solid with five faces, then show their solutions. There are only two ways: 2C or B + C.

 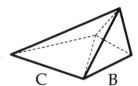

C   C   C   B

Show students how to make a record of a solution.

1. Put a press-on dot on each face.

2. Set the solid on a large piece of paper, removing the press-on dot from the face it is resting on. Carefully trace around that face. Remove a dot from another face and roll it over onto that face and trace it. Continue to roll and trace until all the dots have been removed and all the faces traced. The dots help students keep track of which faces they have traced.

3. Straighten the lines with a ruler. The completed pattern is called a net. Depending where they started to roll and trace, students will have different nets for the same solid. Students may want to find how many different nets can be made from one model.

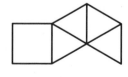   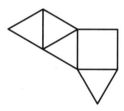

4. Place the pattern on a piece of heavy paper and punch holes with a straight pin to mark each vertex.

5. Use pencil and ruler to connect the holes in the heavy paper. Connect a few lines before showing a completed example prepared before class.

If you are using glue, visualize with your students which pairs of lines will join to form an edge of the model. Put a tab on one side of each pair that will form an edge.

6. Carefully cut the shape out of heavy paper, score the fold lines, and fold. Save demonstration time by cutting another shape out of the heavy paper and partly scoring the lines before the class. Show your students how to score lines with a ball point pen and ruler to make a grove for a crisp fold. After all the lines have been folded, tape or glue the model together.

Exploration

Continue the activity by combining pyramids to find solids with six, seven, eight, nine, and ten faces. Example solutions

six faces 2B or 2A
seven faces A + 2C
eight faces A + 2B, 4C, or 2B + 2C
nine faces 2C + 2A
ten faces 4C + A

Students may make card-stock models to display or they may post the net for each solid to show the number of faces.

Larger combinations of Polyhedra Blocks need to be taped together before they are rolled and traced.

Independent Investigation

Students may want to build complex solids having more than ten faces. Some students may want to use the solids displayed to count and record the number of vertices and edges.

Patterns for Polyhedra Blocks

Block A

Block B

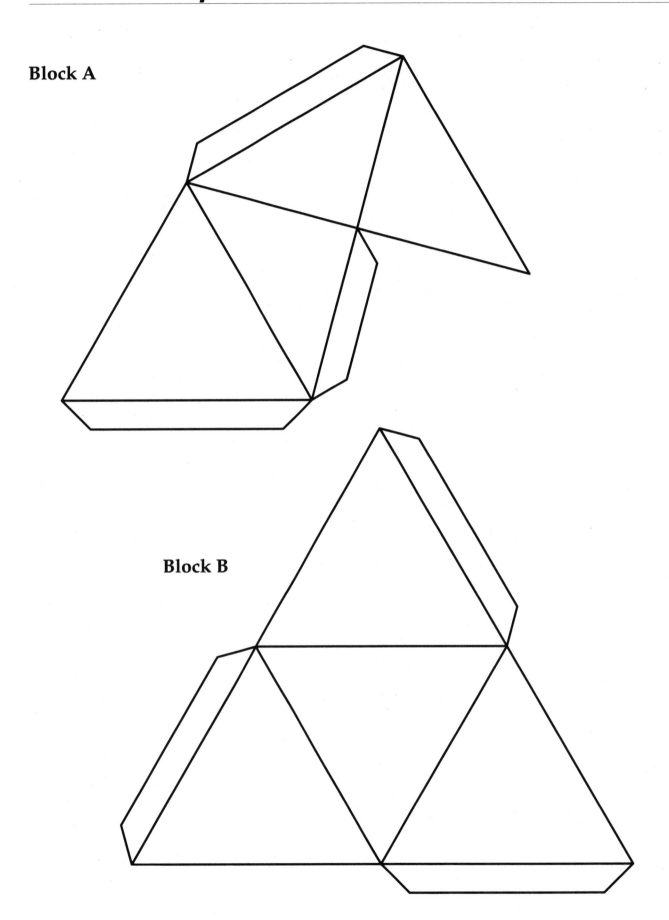

Pattern for Polyhedra Block

Block C

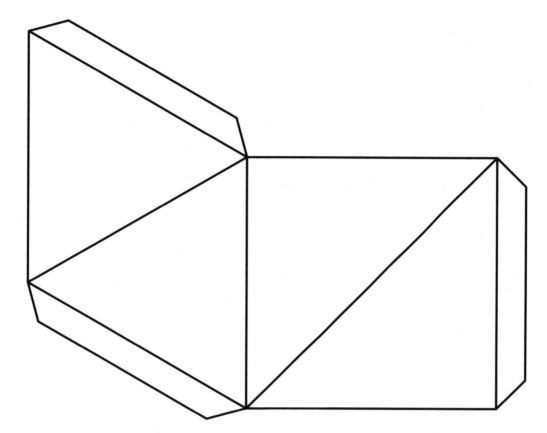

Right Prisms

Materials

Polyhedra Block Right Prisms (page 61)

Make a transparency of the chart or copies for students.

Centimeter grid paper

Copy the grid onto card stock.

Scissors

Transparent tape

Brightly colored press-on dots

Prisms

For younger children, make open containers from transparency film in the shapes of the prisms (page 61) for students to fill.

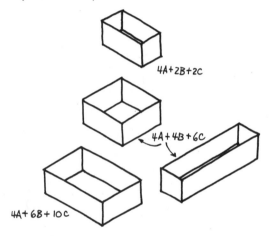

4A+2B+2C

4A+4B+6C

4A+6B+10C

Introduction

Assemble the small cube, using 4 A blocks and 1 B block. Show it to the class, explaining that it is a *right prism*. It is a right prism because all the lateral faces are perpendicular to the bases—each lateral face is a rectangle.

Exploration

Using the Polyhedra Blocks, have students assemble as many right prisms as possible. If you feel frustrated students need a hint, suggest they begin with an A block—it has a square corner to begin building from. Students can construct models of their solutions by measuring the rectangular faces, cutting the correct sizes out of centimeter grid paper, and taping the faces together. Display the models where students can evaluate them. Younger students can fill teacher-made prisms and indicate the blocks used to fill each one.

Closing Discussion

As a class, study the collection of prisms to see whether any are missing. To check whether the solution set is complete, organize the prisms. Usually students will organize them by length. The blackline master on page 61 gives most solutions.

Independent Investigation

Some students may construct additional B and C blocks and explore larger prisms. Prism models could be added to the display.

Polyhedra Block Right Prisms

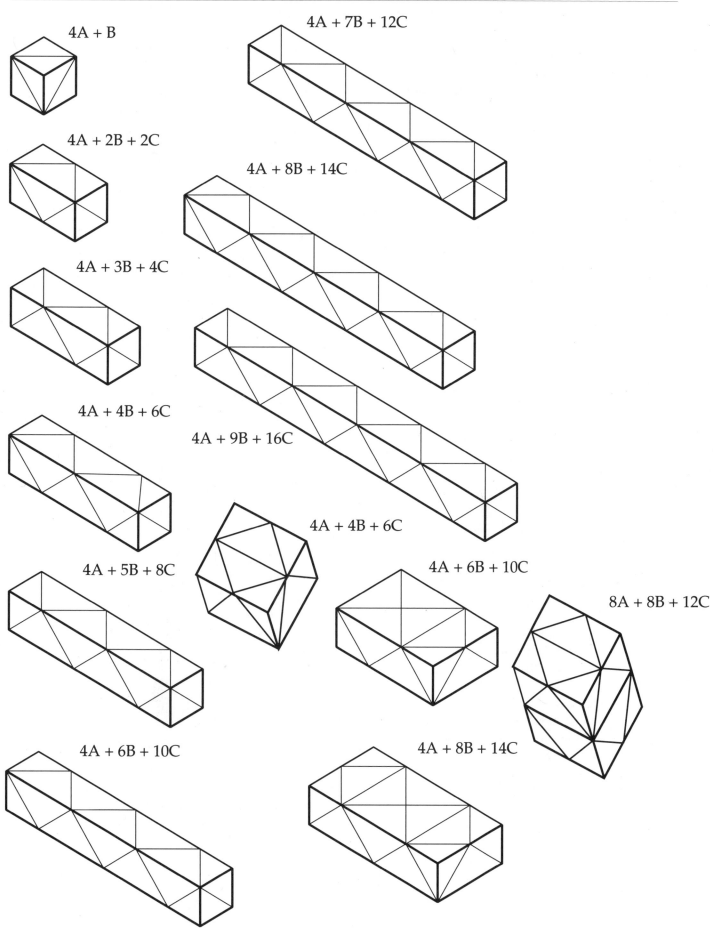

4A + B

4A + 7B + 12C

4A + 2B + 2C

4A + 8B + 14C

4A + 3B + 4C

4A + 4B + 6C

4A + 9B + 16C

4A + 4B + 6C

4A + 5B + 8C

4A + 6B + 10C

8A + 8B + 12C

4A + 6B + 10C

4A + 8B + 14C

Oblique Prisms

Materials
Polyhedra Block Oblique Prisms (pages 63–65)
Make a transparency of the chart or copies for students.
Heavy paper
Scissors
Transparent tape
Brightly colored press-on dots

Introduction
Assemble an oblique triangular prism. Use 1 B block and 2 C blocks. Begin with a square pyramid resting on a triangular face.

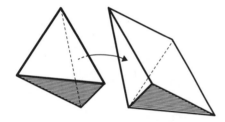

Show it to the class, explaining that it is an *oblique prism* because the lateral faces are not perpendicular to the bases. Each lateral face is a parallelogram.

Exploration
Have the class assemble as many oblique prisms as possible. Use the roll and trace procedures described in the first activity in this exploration, "Comparing Numbers of Faces," to make a net and pattern for card-stock models of these prisms. Display the models for students to evaluate.

Closing Discussion
With your students, study the collection of prisms, to see whether any are missing. It may be difficult to find missing prisms because of the variety of possible bases. Students may identify possible bases that can be formed from the triangular faces on blocks B and C.

Possible bases

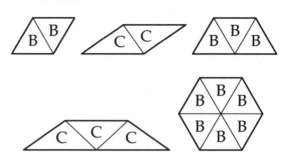

Organize a series of prisms for each possible base to find any missing solids. See the solution page. Once identified, models of the missing prisms can be made to complete the solution set.

In a regular polygon, each side is the same length and all the angles are identical. Look at the triangular prism shown at the beginning of this activity. Are its bases regular or nonregular? (Regular) The class can sort the prisms into two groups—one with regular bases and another with nonregular bases. Discuss which group the hexagonal box kite belongs to. (It is a regular oblique hexagonal prism.)

Independent Investigation
Students may investigate the volumes of the prisms. Can they find an oblique prism that has the same volume as a right prism. Ask them to explain how they know the volumes are the same. These two prisms have the same volume.

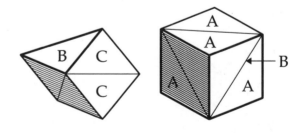

Encourage students to make a prism whose side lengths on the bases are twice that of the oblique triangular prism made with 1 B block and 2 C blocks. Compare its volume to the demonstration prism that is similar to it.

Polyhedra Block Oblique Triangular Prisms

Equilateral Bases

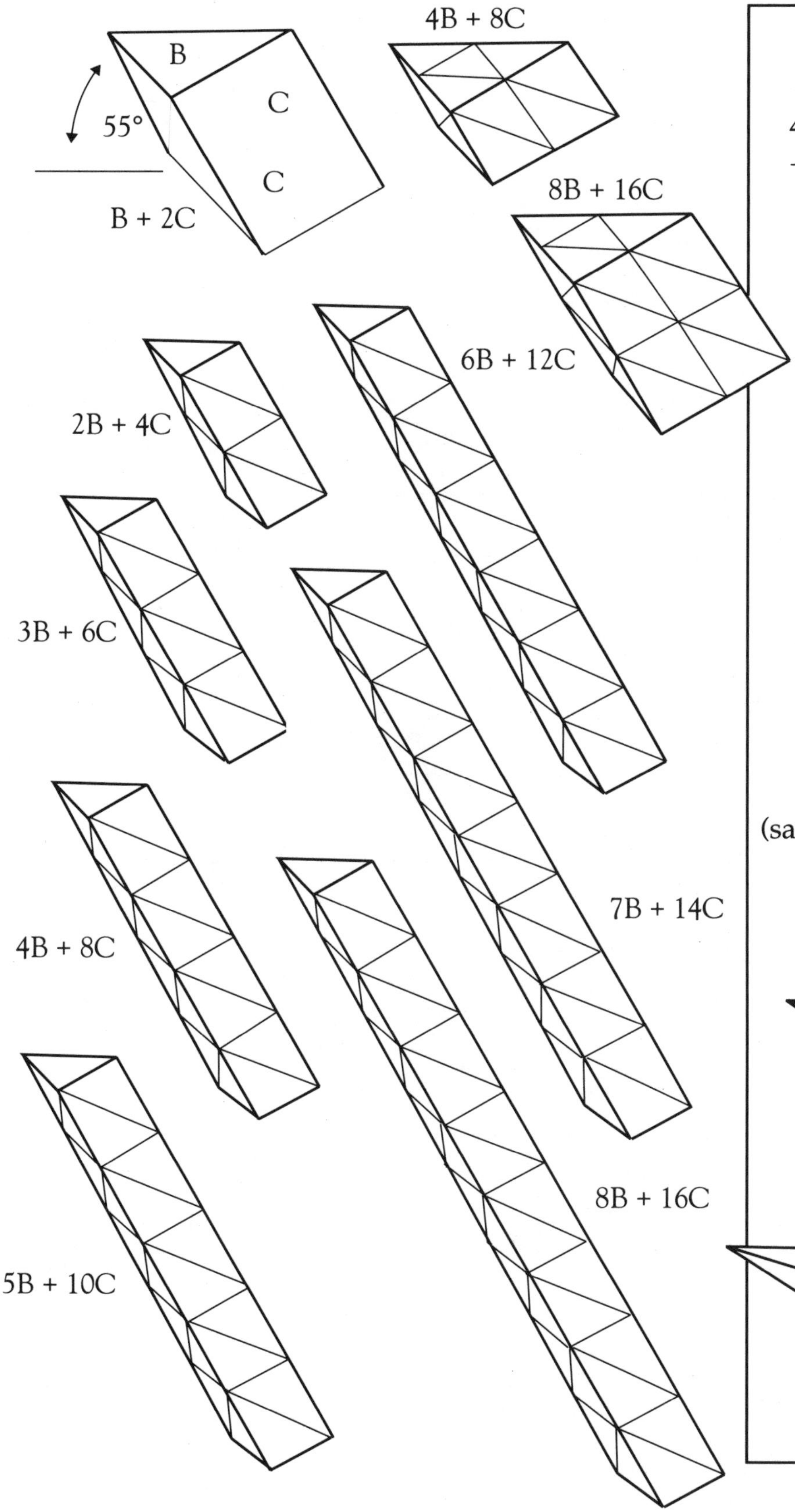

B

55°

B + 2C

B

C

C

4B + 8C

8B + 16C

6B + 12C

2B + 4C

3B + 6C

4B + 8C

5B + 10C

7B + 14C

8B + 16C

Right Isosceles Bases

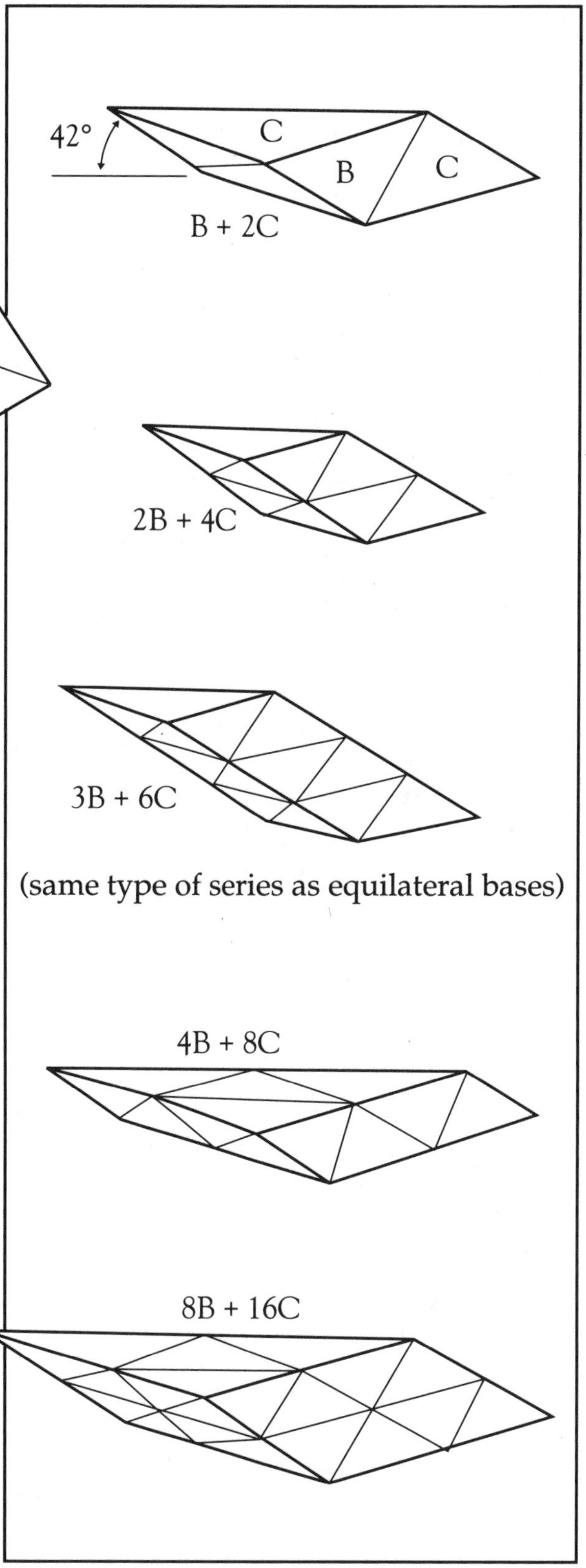

42°

C

B

C

B + 2C

2B + 4C

3B + 6C

(same type of series as equilateral bases)

4B + 8C

8B + 16C

Polyhedra Block Oblique Parallelogram Prisms

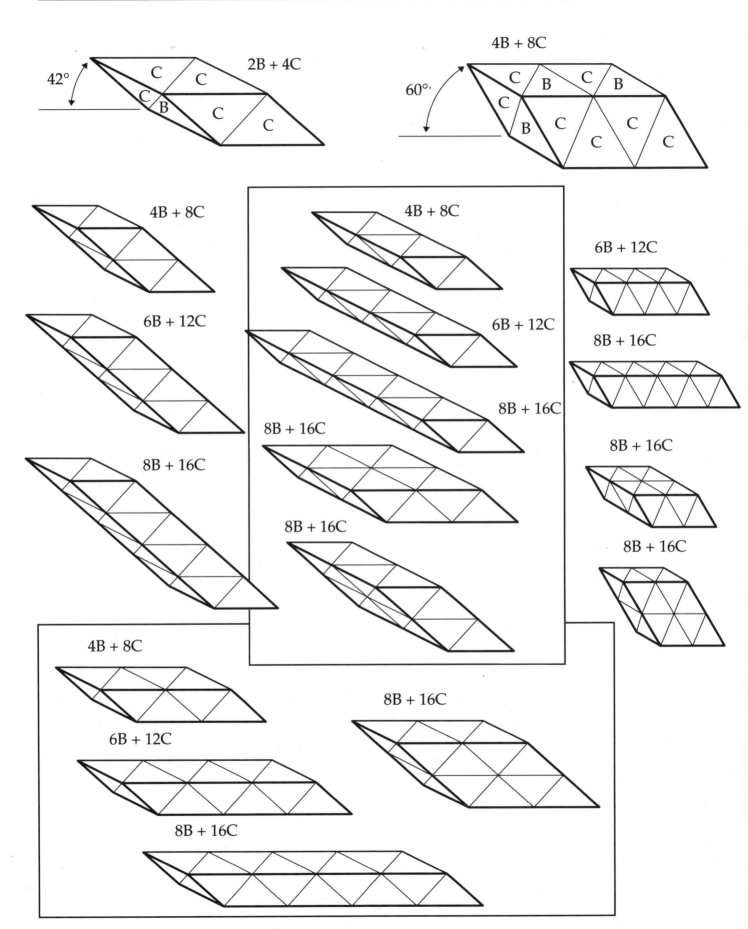

42°   C   C   C   B   C   C   2B + 4C

4B + 8C   60°   C   B   C   B   C   B   C   C   C

4B + 8C

4B + 8C

6B + 12C

6B + 12C

6B + 12C

8B + 16C

8B + 16C

8B + 16C

8B + 16C

8B + 16C

8B + 16C

8B + 16C

4B + 8C

6B + 12C

8B + 16C

8B + 16C

Polyhedra Block Oblique Quadrilateral Prisms

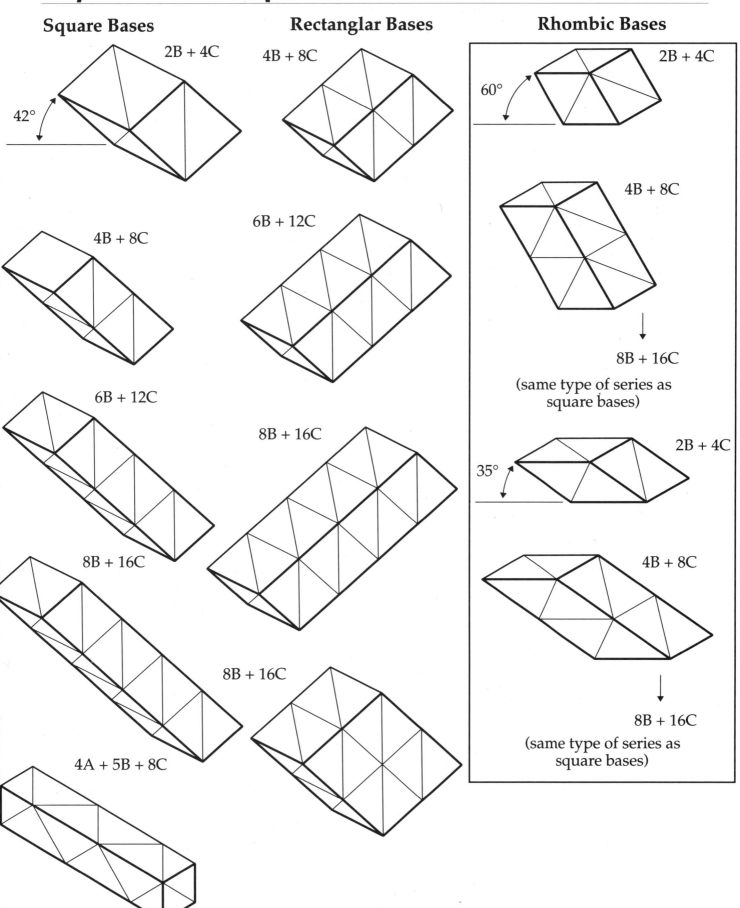

Square Bases

2B + 4C

42°

4B + 8C

6B + 12C

8B + 16C

4A + 5B + 8C

Rectanglar Bases

4B + 8C

6B + 12C

8B + 16C

8B + 16C

Rhombic Bases

60°

2B + 4C

4B + 8C

↓

8B + 16C

(same type of series as square bases)

35°

2B + 4C

4B + 8C

↓

8B + 16C

(same type of series as square bases)

Polyhedra Block Oblique Prisms

Trapezoid Bases

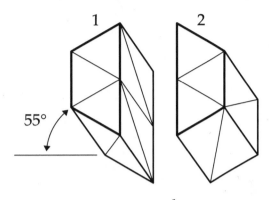

55°

each
3B + 6C

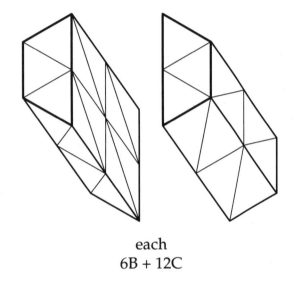

each
6B + 12C

each
3B + 6C

42°

1

2

Regular Hexagon Bases

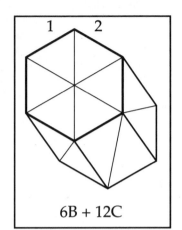

6B + 12C

Nonregular Hexagon Bases

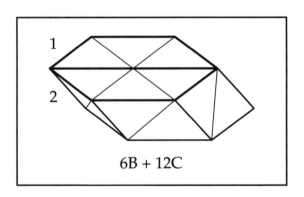

1

2

6B + 12C

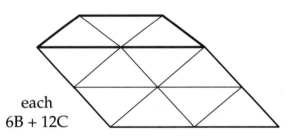

each
6B + 12C

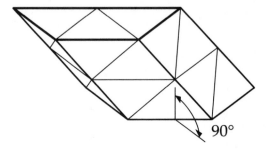

90°

Similar Solids

Materials

Large model similar to the A block made with 1 B
block and 3 C blocks

Assemble this model before class.

Rulers
Card stock or large sheets of tagboard
Scissors
Transparent tape
Brightly colored press-on dots

Introduction

Show an A block and the larger similar pyramid.
Explain that the two pyramids are similar because
their bases are similar triangles (triangles that are the
same shape), and their altitudes have the same ratio
as any two corresponding edges of their bases. In
this case, the ratio of the edges is 2 to 1. Use a ruler to
measure the heights of the two pyramids to show
that the altitude ratio is also 2 to 1.

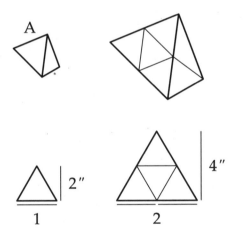

Exploration

Have the class use the Polyhedra Blocks to make
similar, larger models of the B and C blocks. Allow
students to find the ratios for the similar pyramids.
While most students will find similar pyramids
with ratios of 2 to 1, some will find larger pyramids
with ratios of 3 to 1. (See solution page.) Students
may build larger similar solids of some of the
polyhedra they studied in earlier activities. The
square pyramid, the hexahedron, and the small
cube would be good ones to explore.

Students may be challenged to create their own
pairs of similar solids. Using nets explained in
"Comparing Numbers of Faces," students
may make a record of their solutions by
rolling and tracing their solids and
constructing card-stock models for a collection
table.

Closing Discussion

If students made models of similar pairs of solids,
have the class evaluate them. Find out whether each
pair is truly similar by measuring and checking for
correct ratios. If models were not made, allow
several students to take turns rebuilding pairs of
similar solids to show the class.

Compare any series of three similar solids. For
example:

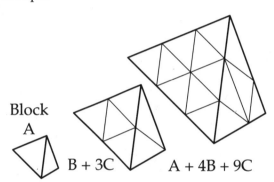

Examine these to see how the ratios changed (from
2 to 1 to 3 to 1). Find the ratio if a fourth, larger
similar pyramid were build, then predict how many
of each block would be needed to make that
pyramid.

Independent Investigation

Students can find the ratios of volume or surface
area for similar solids, and compare them to the
ratios of side length.

Similar Solids

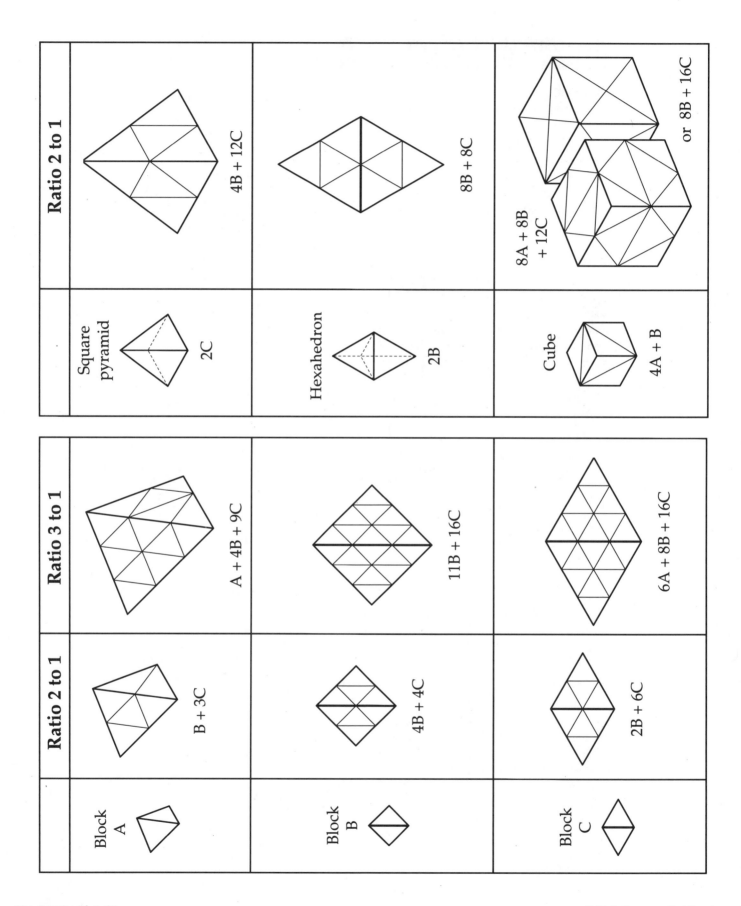

Ratio 2 to 1

Square pyramid — 2C

4B + 12C

Hexahedron — 2B

8B + 8C

Cube — 4A + B

8A + 8B + 12C   or 8B + 16C

Block A

Ratio 2 to 1 — B + 3C

Ratio 3 to 1 — A + 4B + 9C

Block B

Ratio 2 to 1 — 4B + 4C

Ratio 3 to 1 — 11B + 16C

Block C

Ratio 2 to 1 — 2B + 6C

Ratio 3 to 1 — 6A + 8B + 16C

Truncations

Materials
Small cube made with 1 B block and 4 A blocks

Introduction
Explain that *truncate* means to cut off. A *truncated solid* has some or all of its corners cut off.

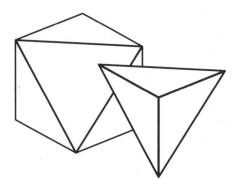

Show the small cube and remove one of the A blocks from a corner. Ask the students what would remain if you truncated three other corners. (A tetrahedron)

Exploration
Have students make several regular polyhedra, truncate the corners, and identify the remaining solids. For example, if each vertex of a cube is truncated by cutting at the middle of each edge, a cuboctahedron is formed. If each vertex of a tetrahedron is truncated making the cut at the middle of each edge, an octahedron is formed. Encourage students to make other solids and truncate them, identifying the resulting solid if possible. Students may make pyramids and truncate the vertex. If they don't, suggest it. Pages 71 and 72 give further ideas for truncation.

Students may use the roll and trace method to make a card-stock model of the truncated shape and of the shape that was cut off. Use double-sided tape to put the two parts together; they may be taken apart to show truncation.

Closing Discussion
Talk about the names of solids that were truncated and the solids formed by that truncation. For example, a truncated cube can be a tetrahedron or a cuboctahedron.

Give students an opportunity to demonstrate their results. Some students may have their own card-stock models to show the class, while others can build a solid with the Polyhedra Blocks and truncate it for the class.

Some of the truncated solids can be organized into a series to show a pattern. The pattern established with the Polyhedra Blocks can be extended to make predictions about larger solids.

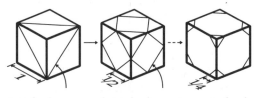

tetrahedron   cuboctahedron   truncated cube

Have students describe the predicted truncated cube. What shapes are its faces? (Octagons, triangles) How many of each kind of face? (6, 8) Does this solid have another name like the tetrahedron or cuboctahedron?

When students show the truncation of a pyramid's vertex, ask them to name the fractional part that was removed. The square pyramid shown on page 70 is made of sixteen blocks, 4 B and 12 C, of equal volume. (The B and C blocks have the same volume, twice that of the A block.) The truncated vertex is made of 2 C blocks. In this case, the truncated part is $\frac{2}{16}$ or $\frac{1}{8}$ of the original square pyramid.

The remaining $\frac{7}{8}$ of the truncated pyramid is called a frustum. A frustum is that part of a pyramid between two parallel planes, in this case the plane of the pyramid's base and the plane of truncation.

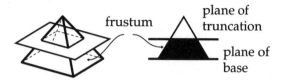

frustum | plane of truncation | plane of base

Independent Investigation

Some students may explore the pieces remaining after truncation: Are there any pieces that are not pyramids? Which solids would produce truncated pieces that are square pyramids? (Octahedron, for example) Which regular solid could be made from all eight truncated corners from a cube? (Octahedron)

Students may build a paper model of the truncated cube explored in the closing discussion.

Students may find the volumes of several frustums. They need to find the volume of the truncated part, then subtract it from the volume of the original pyramid.

Where would a square pyramid be truncated so that the volume of the truncated part was equal to the volume of the pyramid's frustum?

Frustums are a subset of a group of solids called prismoids, nonregular solids whose volume fits entirely between two parallel planes.

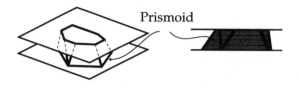

Prismoid

The modified raft in Exploration 5 is an example of a prismoid. See page 31.

Students may make prismoids with Polyhedra Blocks and show them to the class. If models of prismoids are constructed, display them. Ask students to imagine what the possible vertex of each prismoid might look like.

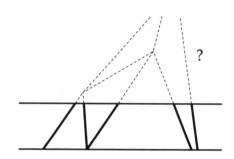

Truncated Solids

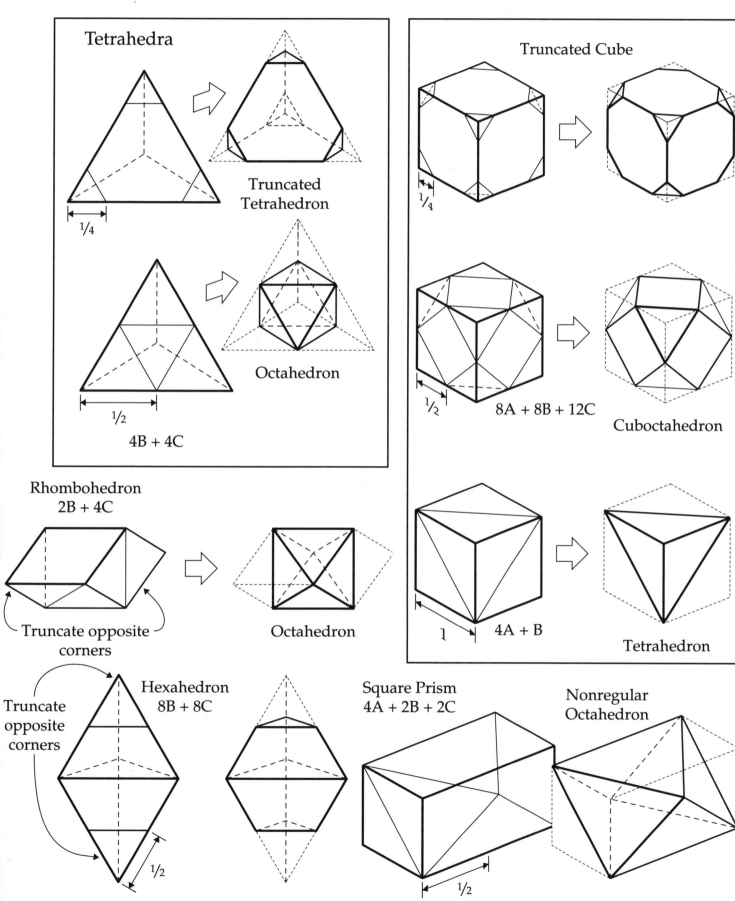

Tetrahedra

Truncated Tetrahedron

$\frac{1}{4}$

Octahedron

$\frac{1}{2}$

$4B + 4C$

Rhombohedron
$2B + 4C$

Truncate opposite corners

Octahedron

Truncate opposite corners

Hexahedron
$8B + 8C$

$\frac{1}{2}$

Truncated Cube

$\frac{1}{4}$

$8A + 8B + 12C$

Cuboctahedron

$\frac{1}{2}$

1 $4A + B$

Tetrahedron

Square Prism
$4A + 2B + 2C$

Nonregular Octahedron

$\frac{1}{2}$

Frustums and Prismoids

Frustums

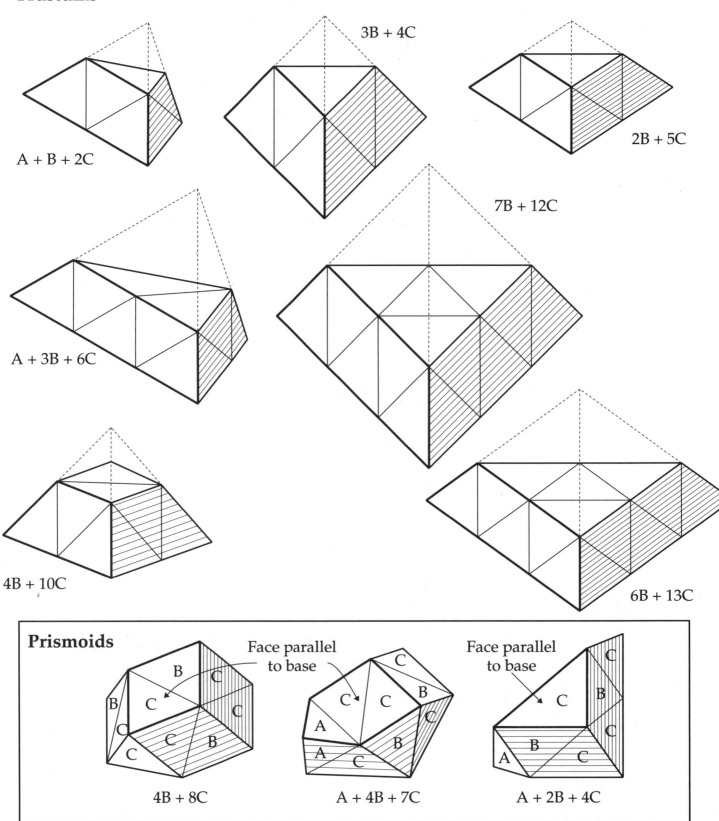

$A + B + 2C$

$3B + 4C$

$2B + 5C$

$A + 3B + 6C$

$7B + 12C$

$4B + 10C$

$6B + 13C$

Prismoids

Face parallel to base

$4B + 8C$

Face parallel to base

$A + 4B + 7C$

Face parallel to base

$A + 2B + 4C$

Intersections of Planes with Solids

Materials

Windows cut in 5" × 8" index cards (patterns on
   page 76)
Scissors or razor knife
Intersections Recording Sheet
   Copy this for your students.

Introduction

If you built the raft variation of the tetrahedral kite in
Exploration 5, display it as an example of an
inclined intersection, or slice, of a tetrahedron.
Demonstrate the intersecting plane by building a
tetrahedron from the Polyhedra Blocks and
inserting a small, thin piece of paper as shown in the
illustration. Trace around the intersected portion of
the tetrahedron, and remove the paper to show the
square you traced. Explain that the square is the
actual area of intersection between the tetrahedron
and the plane.

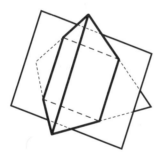

Show how to use the index card windows by
holding up window number one and fitting it over
block B. Explain that this shows that when a plane
and tetrahedron intersect, the intersection can be an
equilateral triangle.

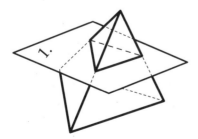

Remove the B block from the top of the large
tetrahedron used earlier and insert a piece of paper,
replacing the B block. Trace around it and remove
the paper to show the class a similar equilateral
triangle.

Exploration

Have students use the windows to make
intersections with blocks A, B, and C. Students will
soon find that the three blocks will only fit in
windows one and two. To find solids that fit the
other windows, they need to combine blocks.

Show students how to draw the intersections they
find on the recording sheet. Write the number of the
window used for the intersection next to the
drawing of it for reference.

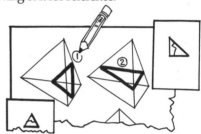

Students may invent their own windows as new
possibilities for intersections occur to them. Have
students cut those windows in 5" × 8" index cards.
Students should experiment with the shape of their
window until it fits over the solid accurately. Be sure
to check the windows students invented. Then add
the new windows to the original set and ask other
students to make solids that fit the new windows.

Examples of possible new windows:

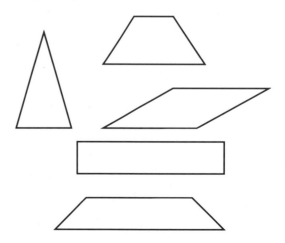

Closing Discussion

Ask your students questions about their exploration. For example,

- Which block, A, B, or C, fits the fewest windows? (Block B is the only regular tetrahedron. A regular solid has only one unique face rather than faces presenting different conditions.)

- Why did the three blocks fit only windows one and two? (Students may guess that since the blocks are made of triangles they will only fit the triangular windows. Remind them of the tetrahedron with the intersecting inclined plane whose intersection was a square. Because of the shape of a tetrahedron, the square window cannot be placed over the pyramid's wide corners.)

- What combinations of blocks fit windows three, four, five, and six? (The combination of blocks B and C fit window three. Because there is a B + C combination in the nonregular octahedron, it will also fit window three. The square pyramid, 4A or 2C, and the regular octahedron, 4C, fit window four. The regular octahedron fits window five. The irregular octahedron fits window six.)

Describe the planes used to cut the polyhedra—if an intersecting plane is parallel to the base of the solid, it is a *horizontal plane*. If it is perpendicular to the base, it is a *vertical plane*. All other intersecting planes are *inclined planes*. Ask your students to use their recording sheets to determine how many of each kind of plane they have pictured.

inclined   horizontal

If the class did not use the recording sheets, have students take turns demonstrating their results by rebuilding solids and fitting the windows on them.

Students who invented their own windows could prove the completed intersections by rebuilding the solid and fitting their window over it.

Independent Investigation

A hexagonal window was not used in this activity. Some students may enjoy finding a solid that would fit a hexagonal window.

The intersection of a plane with the regular tetrahedron in the introduction produced a square. Some students may build larger models of blocks A and C and use the models to determine similar intersections. To produce a square or rectangle, the plane must pass through the base and all three lateral faces.

Intersection Recording Sheet

Name _____

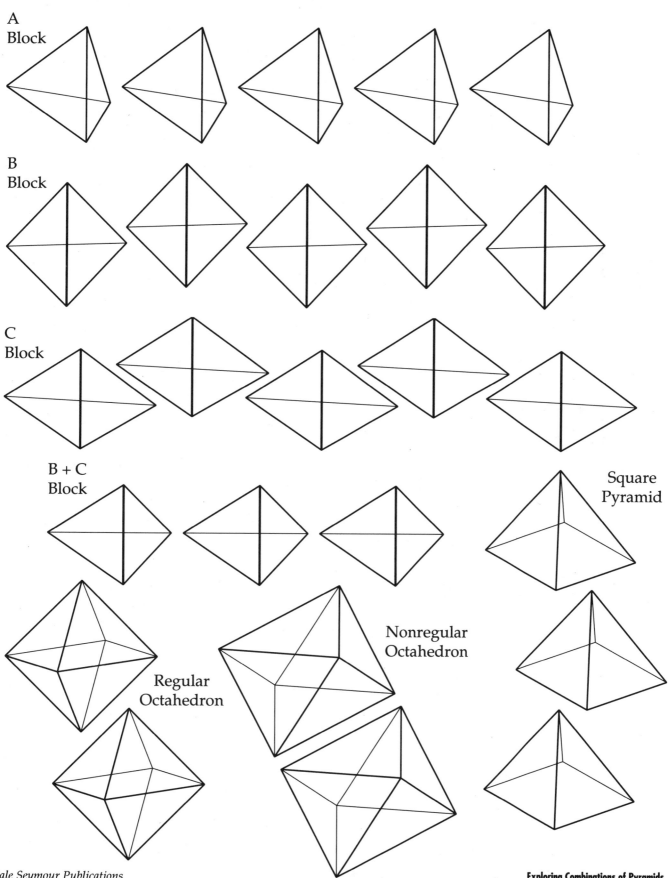

A
Block

B
Block

C
Block

B + C
Block

Square
Pyramid

Regular
Octahedron

Nonregular
Octahedron

Patterns for Windows

Trace each of these polygons onto the middle of a 5" × 8" index card and carefully cut them out with a razor blade knife.

1.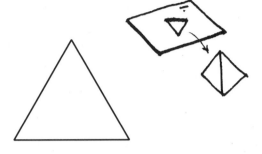

4.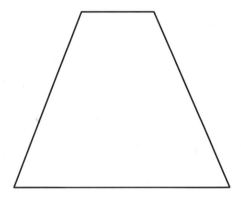

2.

5.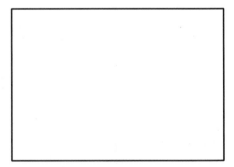

3.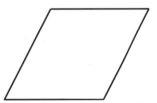

6.

Planes as Bisectors

Materials

Large tetrahedron (4 B blocks and 4 C blocks)
Cuboctahedron (8 B blocks and 12 C blocks)
5" × 8" index cards or pieces of heavy paper
Cuboctahedron Bisectors (page 79)
Copy this for your students.

Introduction

Carefully slide the tetrahedron model apart to form two equal solids, and show both halves.

Reassemble the tetrahedron with a piece of paper or index card between the two halves. Explain that the piece of paper represents a plane that divides the tetrahedron into two equal halves; the plane is a bisector.

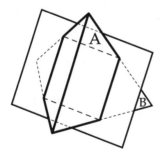

Exploration

Show a model of the cuboctahedron. Explain that they are going to demonstrate a bisector by building half a cuboctahedron on the Cuboctahedron Bisectors pattern sheet. If students have not constructed the cuboctahedron before, tell them that for half of a cuboctahedron, they need 4 B blocks and 6 C blocks.

After a student completes either pattern, have them take the blocks apart and reassemble the half cuboctahedron on the other pattern. Each time,

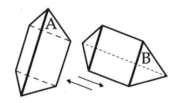

Bisector A intersects the cuboctahedron at a different angle than bisector B, but this dual construction proves that the results are equivalent.

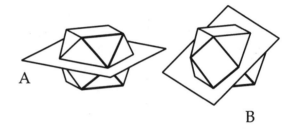

This would be a good place to review angle of intersection from the previous activity.

The tetrahedral pattern of part B may remind some students of one of the kites constructed in Exploration 5.

Closing Discussion

Show the model of the whole cuboctahedron. Ask students to demonstrate where the cuboctahedron could be bisected by separating it into two equal halves.

Suggest that the one of the bisections could again be bisected, yielding $\frac{1}{4}$ of a cuboctahedron. To do this, replace the C blocks on two sides with 2 A blocks each.

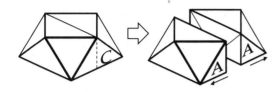

The cuboctahedral bisections and the $\frac{1}{4}$ cuboctahedron, are *prismoids*.

Independent Investigation

Students may find other solids to bisect. For example:

An octahedron made of 4 C blocks can be bisected as shown.

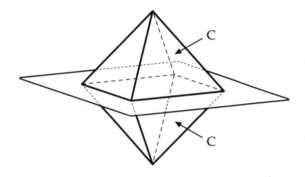

If an octahedron is made from 8 A blocks, it can be bisected in the middle of an edge.

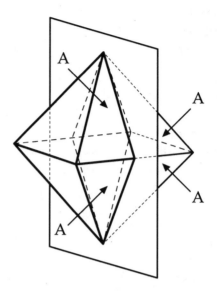

Provide an opportunity for students to share with the class additional examples of bisecting planes.

Cuboctahedron Bisectors

Name _____

Build half of a cuboctahedron on one of these bisectors, then use the same blocks to rebuild half of a cuboctahedron on the other bisector.

Draw in the top view of each construction.

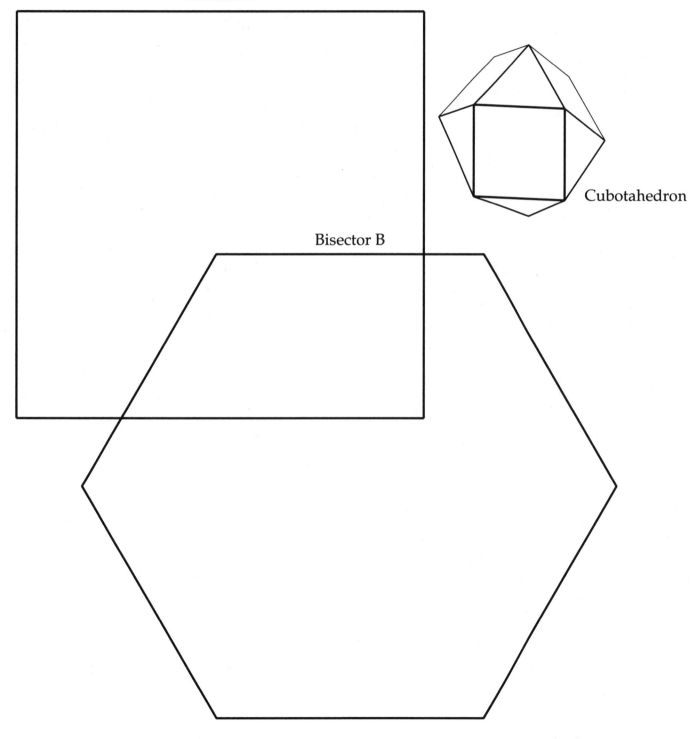

Bisector A

Bisector B

Cubotahedron

Building on Bases

Materials

Trace the right triangular base of a C block on an index card.

Write "Pyramid" on the card.

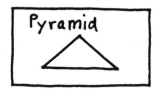

Copies of Base Puzzles (pages 81–84)

Introduction

Explain that students will solve a series of puzzles. The bases of some solids have been traced before the solids were taken apart. The challenge is to rebuild the solid that had the outlined base.

Show the index card with the tracing of a C block. Explain that this is the base of a pyramid. Have the students suggest a block or combination of blocks forming a pyramid that has a base like the tracing. (One C block or two A blocks would fit in the tracing.)

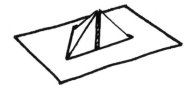

Exploration

Allow the class to explore the pages of traced bases, building solids to fit in them. In several cases, there is more than one way to fill the base. When a solution has been found, take the solid apart and count the number of each type of block used. Then write those numbers inside the shape the solid fits.

Closing Discussion

Students can share solutions for the traced bases by rebuilding the solids that fit them. Often there will be several ways to fill a base. If students just find one solution for those bases, provide time for students to search further.

Independent Investigation

Students may also do the inverse of this activity. Build a solid first, then trace its base to make a puzzle for other students. Make copies of the tracings for other students to explore.

Solutions for Pyramids

page 81
1. A or B
2. 11B + 16C or A + 4B + 9C
3. B + 3C, 4B + 4C, or 4B + 8C (large triangular prism)

page 82
1. 2A or C
2. 6A + 8B + 16C
3. 2B + 6C

page 83
1. 2C + 4A, or 2B + 4C (oblique square prism)
2. 4B + 12C
3. 8A + 8B + 12C or 8B + 16C
4. 4A + B

page 84
1. B + C or 2B + 4C (rhombohedron)
2. 6B + 10C

Base Puzzles Pyramids with Equilateral Bases

Name _____

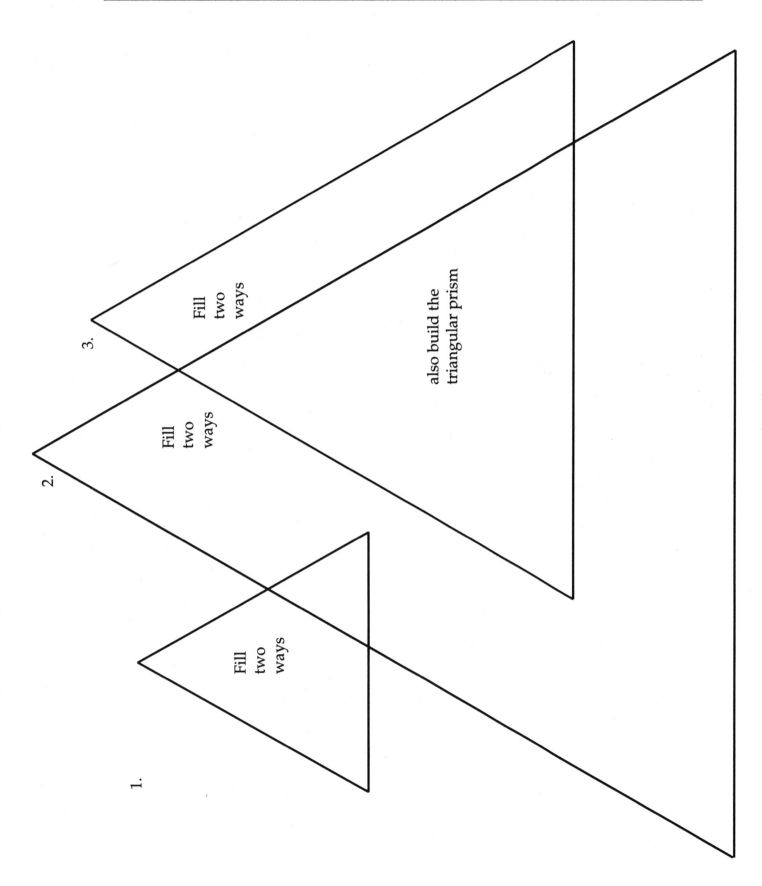

3.

Fill
two
ways

also build the
triangular prism

2.

Fill
two
ways

1.

Fill
two
ways

Name _____

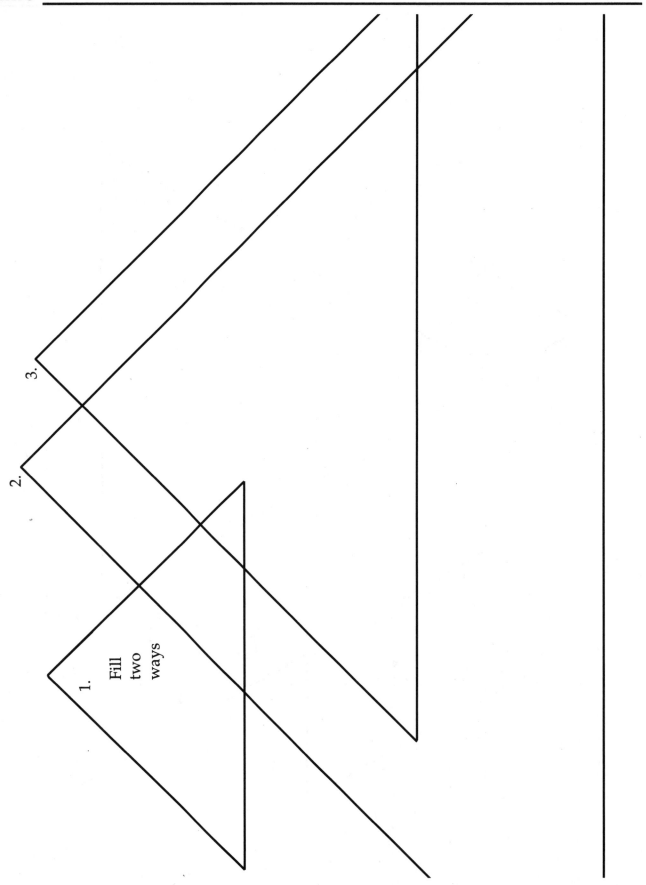

3.

2.

1.

Fill
two
ways

Name _____

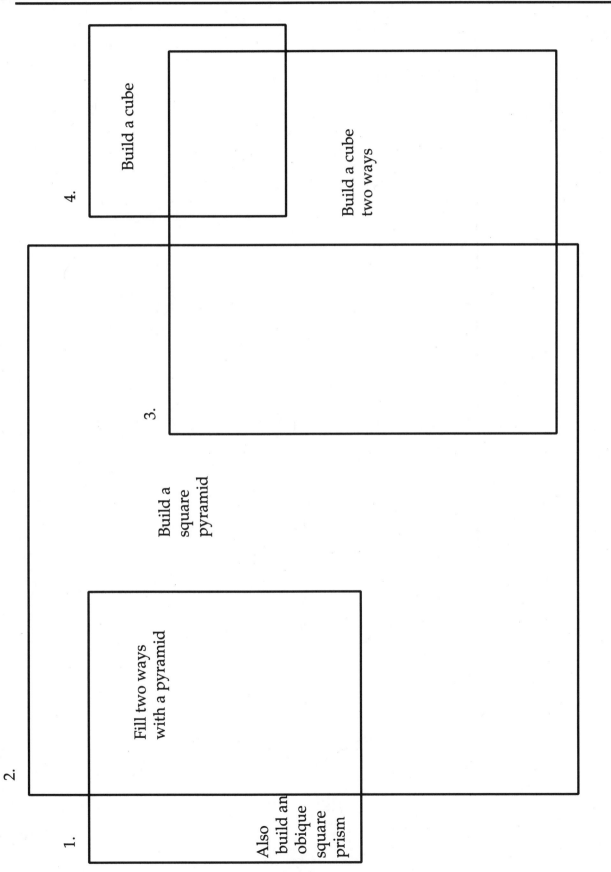

4. Build a cube

Build a cube two ways

3.

Build a square pyramid

2.

Fill two ways with a pyramid

1.

Also build an obique square prism

Base Puzzles Pyramids with Rhombus Bases

Name _____

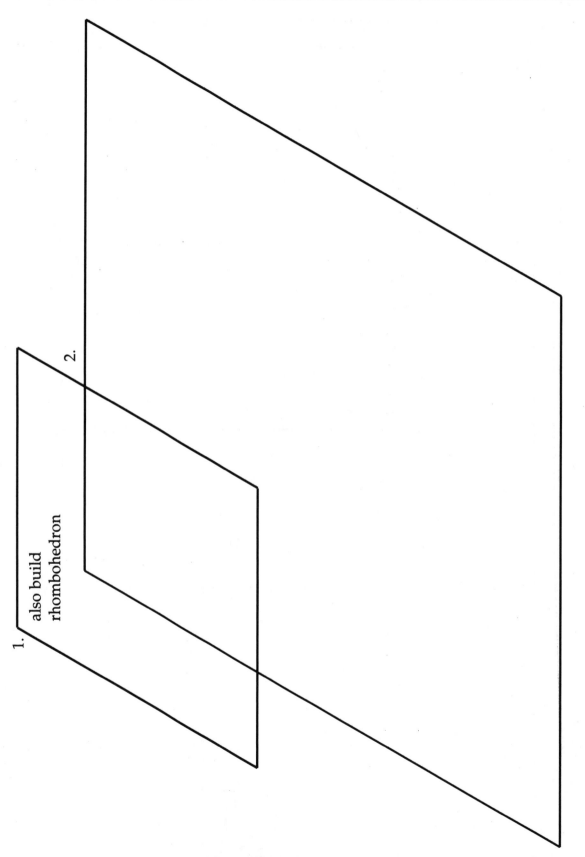

2.

1. also build rhombohedron

VOL. XIV, No. 6   WASHINGTON   JUNE, 1903

THE TETRAHEDRAL PRINCIPLE IN KITE STRUCTURE*

BY ALEXANDER GRAHAM BELL

PRESIDENT OF THE NATIONAL GEOGRAPHIC SOCIETY

IN 1899, at the April meeting, I made a communication to the Academy upon the subject of "Kites with Radial Wings;" and some of the illustrations shown to the Academy at that time were afterwards published in the *Monthly Weather Review*.†

Since then I have been continuously at work upon experiments relating to kites. Why, I do not know, excepting perhaps because of the intimate connection of the subject with the flying-machine problem.

We are all of us interested in aerial locomotion; and I am sure that no one who has observed with attention the flight of birds can doubt for one moment the possibility of aerial flight by bodies specifically heavier than the air. In the words of an old writer, "We cannot consider as impossible that which has already been accomplished."

I have had the feeling that a properly constructed flying-machine should be capable of being flown as a kite; and, conversely, that a properly constructed kite should be capable of use as a flying-machine when driven by its own propellers. I am not so sure, however, of the truth of the former proposition as I am of the latter.

Given a kite, so shaped as to be suitable for the body of a flying-machine, and so efficient that it will fly well in a good breeze (say 20 miles an hour) when loaded with a weight equivalent to that of a man and engine; then it seems to me that this same kite, provided with an actual engine and man in place of the load, and driven by its own propellers at the rate of 20 miles an hour, should be sustained in calm air as a flying-machine. So far as the pressure of the air is concerned, it is surely immaterial whether the air moves against the kite, or the kite against the air.

*A communication made to the National Academy of Sciences in Washington, D. C, April 23, 1903, revised for publication in the NATIONAL GEOGRAPHIC MAGAZINE.
† See *Monthly Weather Review*, April, 1899, vol. xxvii, pp. 154–155, and plate xi

Of course in other respects the two cases are not identical. A kite sustained by a 20-mile breeze possesses no momentum, or rather its momentum is equal to zero, because it is stationary in the air and has no motion proper of its own; but the momentum of a heavy body propelled at 20 miles an hour through still air is very considerable. Momentum certainly aids flight, and it may even be a source of support against gravity quite independently of the pressure of the air. It is perfectly possible, therefore, that an apparatus may prove to be efficient as a flying-machine which cannot be flown as a kite on account of the absence of *vis viva*.

However this may be, the applicability of kite experiments to the flying-machine problem has for a long time past been the guiding thought in my researches.

I have not cared to ascertain how high a kite may be flown or to make one fly at any very great altitude. The point I have had specially in mind is this: That the equilibrium of the structure in the air should be perfect; that the kite should fly steadily, and not move about from side to side or dive suddenly when struck by a squall, and that when released it should drop slowly and gently to the ground without material oscillation. I have also considered it important that the framework should possess great strength with little weight.

I believe that in the form of structure now attained the properties of strength, lightness, and steady flight have been united in a remarkable degree.

In my younger days the word "kite" suggested a structure of wood in the form of a cross covered with paper forming a diamond-shaped surface longer one way than the other, and provided with a long tail composed of a string with numerous pieces of paper tied at intervals upon it. Such a kite is simply a toy. In Europe and America, where kites of this type prevailed, kite-flying was pursued only as an amusement for children, and the improvement of the form of structure was hardly considered a suitable subject of thought for a scientific man.

In Asia kite-flying has been for centuries the amusement of adults, and the Chinese, Japanese, and Malays have developed tailless kites very much superior to any form of kite known to us until quite recently.

It is only within the last few years that improvements in kite structure have been seriously considered, and the recent developments in the art have been largely due to the efforts of one man— Mr Laurence Hargrave, of Australia.

Hargrave realized that the structure best adapted for what is called a "good kite" would also be suitable as the basis for the structure of a flying-machine. His researches, published by the Royal Society of New South Wales, have attracted the attention of the world, and form the starting point for modern researches upon the subject in Europe and America.

Anything relating to aerial locomotion has an interest to very many minds, and scientific kite-flying has everywhere been stimulated by Hargrave's experiments.

In America, however, the chief stimulus to scientific kite-flying has been the fact developed by the United States Weather Bureau, that important information could be obtained concerning weather conditions if kites could be constructed capable of lifting meteorological instruments to a great elevation in the free air. Mr Eddy and others in America have taken the Malay tailless kite as a basis for their experiments, but Professor Marvin, of the United States Weather Bureau; Mr Rotch, of the Blue Hill Observatory, and many others have adapted Hargrave's box kite for the purpose.

Congress has made appropriations to the Weather Bureau in aid of its kite experiments, and a number of meteorological stations throughout the United

States were established a few years ago equipped with the Marvin kite.

Continuous meteorological observations at a great elevation have been made at the Blue Hill Observatory in Massachusetts, and Mr Rotch has demonstrated the possibility of towing kites at sea by means of steam vessels so as to secure a continuous line of observations all the way across the Atlantic.

HARGRAVE'S BOX KITE

Hargrave introduced what is known as the "cellular construction of kites." He constructed kites composed of many cells, but found no substantial improvement in many cells over two alone ; and a kite composed of two rectangular cells

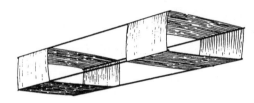

FIG. I—HARGRAVE BOX KITE

separated by a considerable space is now universally known as "the Hargave box kite." This represents, in my opinion, the high-water mark of progress in the nineteenth century ; and this form of kite forms the starting point for my own researches (Fig. 1).

The front and rear cells are connected together by a framework, so that a considerable space is left between them. This space is an essential feature of the kite : upon it depends the fore and aft stability of the kite. The greater the space, the more stable is the equilibrium of the kite in a fore and aft direction, the more it tends to assume a horizontal position in the air, and the less it tends to dive or pitch like a vessel in a rough sea. Pitching motions or oscillations are almost entirely suppressed when the space between the cells is large.

Each cell is provided with vertical sides ; and these again seem to be essential elements of the kite contributing to lateral stability. The greater the extent of the vertical sides, the greater is the stability in the lateral direction, and the less tendency has the kite to roll, or move from side to side, or turn over in the air.

In the foregoing drawing I have shown only necessary details of construction, with just sufficient framework to hold the cells together.

It is obvious that a kite constructed as shown in Fig. 1 is a very flimsy affair. It requires additions to the framework of various sorts to give it sufficient strength to hold the aeroplane surfaces in their proper relative positions and prevent distortion, or bending or twisting of the kite frame under the action of the wind.

Unfortunately the additions required to give rigidity to the framework all detract from the efficiency of the kite : First, by rendering the kite heavier, so that the ratio of weight to surface is increased ; and, secondly, by increasing the head resistance of the kite. The interior bracing advisable in order to preserve the cells from distortion comes in the way of the wind, thus adding to the *drift* of the kite without contributing to the *lift*.

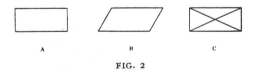

A   B   C

FIG. 2

A rectangular cell like *A* (Fig. 2) is structurally weak, as can readily be demonstrated by the little force required to distort it into the form shown at *B*. In order to remedy this weakness, internal bracing is advisable of the character shown at *C*.

This internal bracing, even if made of the finest wire, so as to be insignificant in weight, all comes in the way of the wind, increasing the head resistance without counterbalancing advantages.

TRIANGULAR CELLS IN KITE CONSTRUCTION

In looking back over the line of experiments in my own laboratory, I recognize that the adoption of a triangular cell was a step in advance, constituting indeed one of the milestones of progress, one of the points that stand out clearly against the hazy background of multitudinous details.

The following (Fig 3) is a drawing of a typical triangular-celled kite made upon the same general model as the Hargrave box kite shown in Fig. 1.

A triangle is by its very structure perfectly braced in its own plane, and in a triangular-celled kite like that shown in Fig. 3, internal bracing of any

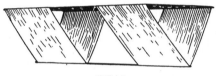

FIG. 3

character is unnecessary to prevent distortion of a kind analogous to that referred to above in the case of the Hargrave rectangular cell (Fig. 2).

The lifting power of such a triangular cell is probably less than that of a rectangular cell, but the enormous gain in structural strength, together with the reduction of head resistance and weight due to the omission of internal bracing, counterbalances any possible deficiency in this respect.

The horizontal surfaces of a kite are those that resist descent under the influence of gravity, and the vertical surfaces prevent it from turning over in the air. Oblique aeroplanes may therefore conveniently be resolved into horizontal and vertical equivalents, that is, into supporting surfaces and steadying surfaces.

The oblique aeroplane A, for example (Fig. 4), may be considered as equivalent in function to the two aeroplanes B and C. The material composing the aeroplane A, however, *weighs less* than the material required to form the two aeroplanes B and C, and the frame-

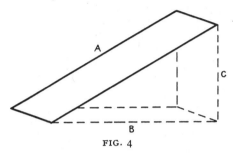

FIG. 4

work required to support the aeroplane A weighs less than the two frameworks required to support B and C.

In the triangular cell shown in Fig. 5, the oblique surfaces ab, bc, are equivalent in function to the three surfaces ad, de, ec, but weigh less. The oblique surfaces are therefore advantageous.

The only disadvantage in the whole arrangement is that the air has not as free access to the upper aeroplane ac, in the triangular form of cell as in the quadrangular form, so that the aeroplane

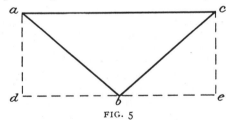

FIG. 5

ac is not as efficient in the former construction as in the latter.

While theoretically the triangular cell is inferior in lifting power to Hargrave's four-sided rectangular cell, practically there is no substantial difference. So far as I can judge from observation in the field, kites constructed on the same general model as the Hargrave Box Kite, but with triangular cells instead of quadrangular, seem to fly as well as the ordinary Hargrave form, and at as high an angle.

Such kites are therefore superior, for they fly substantially as well, while at the same time they are stronger in construction, lighter in weight, and offer less head resistance to the wind.

and B (Fig. 7) may be constructed, as shown at C and D, with advantage, for the weight of the compound kite is thus reduced without loss of structural strength. In this case the weight of the compound kite is *less* than the sum of the weights of the component kites,

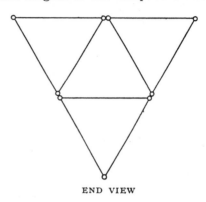

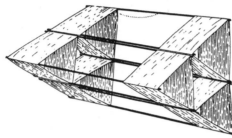

PERSPECTIVE VIEW   END VIEW

FIG. 6—COMPOUND TRIANGULAR KITE

Triangular cells also are admirably adapted for combination into a compound structure, in which the aeroplane surfaces do not interfere with one another. For example, three triangular-celled kites, tied together at the corners, form a compound cellular kite (Fig. 6) which flies perfectly well.

The weight of the compound kite is the sum of the weights of the three kites of which it is composed, and the total aeroplane surface is the sum of the surfaces of the three kites. The ratio of weight to surface therefore is the same in the larger compound kite as in the smaller constituent kites, considered individually.

It is obvious that in compound kites of this character the doubling of the longitudinal sticks where the corners of adjoining kites come together is an unnecessary feature of the combination, for it is easy to construct the compound kite so that one longitudinal stick shall be substituted for the duplicated sticks.

For example: The compound kites A

while the surface remains the same.

If kites could only be successfully compounded in this way indefinitely

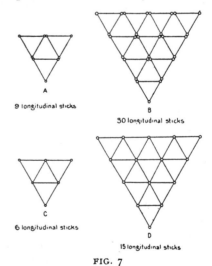

FIG. 7

we would have the curious result that the ratio of weight to surface would diminish with each increase in the size of the compound kite. Unfortunately, however, the conditions of stable flight demand a considerable space between the front and rear sets of cells (see Fig. 6); and if we increase the diameter

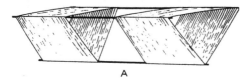

A

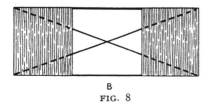

B

FIG. 8

of our compound structure without increasing the length of this space we injure the flying qualities of our kite. But every increase of this space in the fore and aft direction involves a corresponding increase in the length of the empty framework required to span it, thus adding dead load to the kite and increasing the ratio of weight to surface.

the character shown at *B* to prevent distortion under the action of the wind. The necessary bracing, however, not being in the way of the wind, does not materially affect the head resistance of the kite, and is only disadvantageous by adding dead load, thus increasing the ratio of weight to surface.

THE TETRAHEDRAL CONSTRUCTION OF KITES

Passing over in silence multitudinous experiments in kite construction carried on in my Nova Scotia laboratory, I come

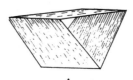

A   B

FIG. 9—A. A TRIANGULAR CELL
B. A WINGED TETRAHEDRAL CELL

to another conspicuous point of advance—another milestone of progress— the adoption of the triangular construction *in every direction* (longitudinally as well as transversely); and the clear realization of the fundamental importance of the skeleton of a tetrahedron, especially the regular tetrahedron, as

Acute-angled tetrahedron   Regular tetrahedron   Right-angled tetrahedron   Obtuse-angled tetrahedron

FIG. 10—WINGED TETRAHEDRAL CELLS

While kites with triangular cells are strong in a transverse direction (from side to side), they are structurally weak in the longitudinal direction (fore and aft), for in this direction the kite frames are rectangular.

Each side of the kite *A*, for example (Fig. 8), requires diagonal bracing of

an element of the structure or framework of a kite or flying-machine.

Consider the case of an ordinary triangular cell *A* (Fig. 9) whose cross-section is triangular laterally, but quadrangular longitudinally.

If now we make the longitudinal as well as transverse cross-sections triangular, we arrive at the form of cell shown at *B*, in which the framework forms the outline of a tetrahedron. In this case the aeroplanes are triangular, and the whole arrangement is strongly suggestive of a pair of birds' wings

FIG. 11—ONE-CELLED TETRAHEDRAL FRAME

raised at an angle and connected together tip to tip by a cross-bar (see *B*, Fig. 9 ; also drawings of winged tetrahedral cells in Fig. 10).

A tetrahedron is a form of solid bounded by four triangular surfaces.

In the regular tetrahedron the boundaries consist of four equilateral triangles and six equal edges. In the skeleton form the edges alone are represented, and the skeleton of a regular tetrahedron is produced by joining together six equal

FIG. 12—FOUR-CELLED TETRAHEDRAL FRAME

rods end to end so as to form four equilateral triangles.

Most of us no doubt are familiar with the common puzzle—how to make four triangles with six matches. Give six matches to a friend and ask him to arrange them so as to form four complete equilateral triangles. The difficulty lies in the unconcious assumption of the experimenter that the four triangles should all be in the same plane. The moment he realizes that they need not be in the same plane the solution of the problem becomes easy. Place three matches on the table so as to form a triangle, and stand the other three up

over this like the three legs of a tripod stand. The matches then form the skeleton of a regular tetrahedron. (See figure 11.)

A framework formed upon this model of six equal rods fastened together at the ends constitutes a tetrahedral cell possessing the qualities of strength and lightness in an extraordinary degree.

It is not simply braced in two directions in space like a triangle, but in three directions like a solid. If I may coin a word, it possesses "*three-dimensional*" strength; not "two-dimensional" strength like a triangle, or "one-dimensional" strength like a rod. It is the skeleton of a solid, not of a surface or a line.

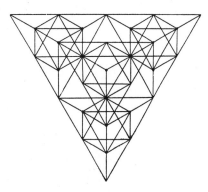

FIG. 13—SIXTEEN-CELLED TETRAHEDRAL FRAME

It is astonishing how solid such a framework appears even when composed of very light and fragile material ; and compound structures formed by fastening these tetrahedral frames together at the corners so as to form the skeleton of a regular tetrahedron on a larger scale possess equal solidity.

Figure 12 shows a structure composed of four frames like figure 11, and figure 13 a structure of four frames like figure 12.

When a tetrahedral frame is provided with aero-surfaces of silk or other material suitably arranged, it becomes a tetrahedral kite, or kite having the form of a tetrahedron.

The kite shown in figure 14 is composed of four winged cells of the regular tetrahedron variety (see Fig. 10), connected together at the corners. Four kites like figure 14 are combined in figure 15, and four kites like figure 15 in figure 16 (at *D*).

Upon this mode of construction an empty space of octahedral form is left in the middle of the kite, which seems to have the same function as the space between the two cells of the Hargrave box kite. The tetrahedral kites that have the largest central spaces preserve their equilibrium best in the air.

reason why this principle of combination should not be applied indefinitely so as to form still greater combinations.

The weight relatively to the wing-surface remains the same, however large the compound kite may be.

The four-celled kite *B*, for example, weighs four times as much as one cell and has four times as much wing-surface, the 16-celled kite *C* has sixteen times as much weight and sixteen times as much-wing surface, and the 64-celled kite *D* has sixty-four times as much weight and sixty-four times as much wing-surface. The ratio of weight to

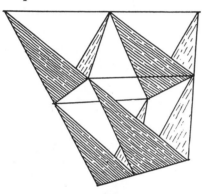

FIG. 14—FOUR-CELLED TETRAHEDRAL KITE

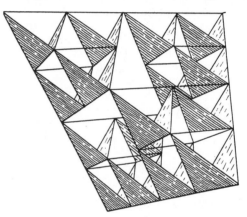

FIG. 15—SIXTEEN-CELLED TETRAHEDRAL KITE

The most convenient place for the attachment of the flying cord is the extreme point of the bow. If the cord is attached to points successively further back on the keel, the flying cord makes a greater and greater angle with the horizon, and the kite flies more nearly overhead; but it is not advisable to carry the point of attachment as far back as the middle of the keel. A good place for high flights is a point half-way between the bow and the middle of the keel.

In the tetrahedral kites shown in the plate (Fig. 16) the compound structure has itself in each case the form of the regular tetrahedron, and there is no

surface, therefore, is the same for the larger kites as for the smaller.

This, at first sight, appears to be somewhat inconsistent with certain mathematical conclusions announced by Prof. Simon Newcomb in an article entitled "Is the Air-ship Coming," published in *McClure's Magazine* for September, 1901—conclusions which led him to believe that "the construction of an aerial vehicle which could carry even a single man from place to place at pleasure requires the discovery of some new metal or some new force."

The process of reasoning by which Professor Newcomb arrived at this re-

Tetrahedral Principle in Kite Structure

FIG. 16—TETRAHEDRAL KITES

A. A WINGED TETRAHEDRAL CELL.   B. A FOUR-CELLED TETRAHEDRAL KITE
C. A SIXTEEN-CELLED TETRAHEDRAL KITE   D. A SIXTY-FOUR-CELLED TETRAHEDRAL KITE

markable result is undoubtedly correct. His conclusion, however, is open to question, because he has drawn a general conclusion from restricted premises. He says:

"Let us make two flying-machines exactly alike, only make one on double the scale of the other in all its dimensions. We all know that the volume, and therefore the weight, of two similar bodies are proportional to the cubes of their dimensions. The cube of two is eight: hence the large machine will have eight times the weight of the other. But surfaces are as the squares of the dimensions. The square of two is four. The heavier machine will therefore expose only four times the wing surface to the air, and so will have a distinct disadvantage in the ratio of efficiency to weight."

a giant kite that should lift a man— upon the model of the Hargrave box kite. When the kite was constructed with two cells, each about the size of a small room, it was found that it would take a hurricane to raise it into the air. The kite proved to be not only incompetent to carry a load equivalent to the weight of a man, but it could not even raise *itself* in an ordinary breeze in which smaller kites upon the same model flew perfectly well. I have no doubt that other investigators also have fallen into the error of supposing that large structures would necessarily be capable of flight, because exact models of them,

FIG. 17—THE AERODROME KITE

Professor Newcomb shows that where two flying-machines—or kites, for that matter—are exactly alike, only differing in the scale of their dimensions, the ratio of weight to supporting surface is greater in the larger than the smaller, increasing with each increase of dimensions. From which he concludes that if we make our structure large enough it will be too heavy to fly.

This is certainly true, so far as it goes, and it accounts for my failure to make

made upon a smaller scale, have demonstrated their ability to sustain themselves in the air. Professor Newcomb has certainly conferred a benefit upon investigators by so clearly pointing out the fallacious nature of this assumption.

But Professor Newcomb's results are probably only true when restricted to his premises. For models *exactly alike, only differing in the scale of their dimensions,* his conclusions are undoubtedly sound; but where large kites are formed by the multiplication of smaller kites into a cellular structure the results are very different. My own experiments with compound kites composed of triangular cells connected corner to corner have amply demonstrated the fact that the dimensions of such a kite may be increased to a very considerable extent without materially increasing the ratio of weight to supporting surface; and upon the tetrahedral plan (Fig. 16) the weight relatively to the wing-surface remains the same however large the compound kite may be.

The indefinite expansion of the triangular construction is limited by the fact that dead weight in the form of empty framework is necessary in the central space between the sets of cells (see Fig. 6), so that the necessary increase of this space when the dimensions of the compound kite are materially increased—in order to preserve the stability of the kite in the air—adds still more dead weight to the larger structures. Upon the tetrahedral plan illustrated in Figs. 14, 15, 16, no necessity exists for empty frameworks in the central spaces, for the mode of construction gives solidity without it.

Tetrahedral kites combine in a marked degree the qualities of strength, lightness, and steady flight; but further experiments are required before deciding that this form is the best for a kite, or that winged cells without horizontal aeroplanes constitute the best arrangement of aero-surfaces.

The tetrahedral principle enables us to construct out of light materials solid frameworks of almost any desired form, and the resulting structures are admi-

rably adapted for the support of aero-surfaces of any desired kind, size, or shape (aeroplanes or aerocurves, etc., large or small).

In further illustration of the tetrahedral principle as applied to kite construction, I show in figure 17 a photograph of a kite which is not itself tetrahedral in form, but the framework of which is built up of tetrahedral cells.

This kite, although very different in construction and appearance from the Aerodrome of Professor Langley, which

FIG. 18—THE AERODROME KITE JUST RISING INTO THE AIR WHEN PULLED BY A HORSE

I saw in successful flight over the Potomac a few years ago, has yet a suggestiveness of the Aerodrome about it, and it was indeed Professor Langley's apparatus that led me to the conception of this form.

The wing surfaces consist of horizontal aeroplanes, with oblique steadying surfaces at the extremities. The body of the machine has the form of a boat, and the superstructure forming the support for the aeroplanes extends across the boat on either side at two points near the bow and stern. The aeroplane surfaces form substantially two pairs of wings, arranged dragon-fly fashion.

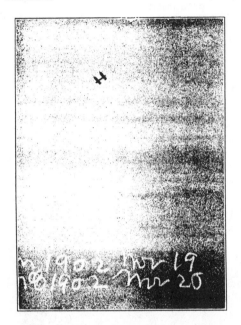

FIG. 19—AERODROME KITE IN THE AIR

The whole framework for the boat and wings is formed of tetrahedral cells having the form of the regular tetrahedron, with the exception of the diagonal bracing at the bottom of the superstructure ; and the kite turns out to be strong, light, and a steady flyer.

I have flown this kite in a calm by attaching the cord—in this case a Manila rope—to a galloping horse. Figure 18 shows a photograph of the kite just rising into the air, with the horse in the foreground, but the connecting rope does not show. Figure 19 is a photograph of the kite at its point of greatest elevation, but the horse does not appear in the picture. Upon releasing the rope the kite descended so gently that no damage was done to the apparatus by contact with the ground.

Figure 20 shows a modified form of the same kite, in which, in addition to the central boat, there were two side floats, thus adapting the whole structure to float upon water without upsetting.

An attempt which almost ended disastrously, was made to fly this kite in a good sailing breeze, but a squall struck it before it was let go. The kite went up, lifting the two men who held it off their feet. Of course they let go instantly, and the kite rose steadily in the air until the flying cord (a Manila rope

FIG 20—FLOATING KITE

⅜ inch diameter) made an angle with the horizon of about 45° when the rope snapped under the strain.

Tremendous oscillations of a pitching character ensued ; but the kite was at such an elevation when the accident happened, that the oscillations had time to die down before the kite reached the ground, when it landed safely upon even keel in an adjoining field and was found to be quite uninjured by its rough experience.

Kites of this type have a much greater lifting power than one would at first sight suppose. The natural assumption is that the winged superstructure alone supports the kite in the air, and that the boat body and floats represent mere dead-load and head resistance. But this is far from being the case. Boat-shaped bodies having a V-shaped cross-section are themselves capable of flight and expose considerable surface to the wind. I have successfully flown a boat of this kind as a kite without any superstructure whatever, and although it did not fly well, it certainly supported itself in the air, thus demonstrating the fact that the boat surface is an element of support in compound structures like those shown in figures 17 and 20.

Of course the use of a tetrahedral cell is not limited to the construction of a framework for kites and flying-machines. It is applicable to any kind of structure whatever in which it is desirable to combine the qualities of strength and lightness. Just as we can build houses of all kinds out of bricks, so we can build structures of all sorts out of tetrahedral frames, and the structures can be so formed as to possess the same qualities of strength and lightness which are characteristic of the individual cells. I have already built a house, a framework for a giant wind-break, three or four boats, as well as several forms of kites, out of these elements.

It is not my object in this communication to describe the experiments that have been made in my Nova Scotia laboratory, but simply to bring to your attention the importance of the tetrahedral principle in kite construction.